Faster Than Light (Gravitic) Propulsion

Via the Quantum Zeno Effect

Dr. William Joseph Bray - FORWARD AND INTRODUCTION

There is 10 times more energy [on the moon] there than there ever was in fossil fuel on the Earth.
--Gerald Kulcinski, nuclear engineering professor

This paper describes and details an engine design that functions by bending space using sound and proven hardened principles in Quantum Physics to produce an Alcubierre Space-time Manifold as described in 1994. The energy requirements are negligible because the process only uses detection and measurement. The entire principle is an overlooked relationship between the Quantum Zeno Effect and General Relativity.

First, it is strongly recommended that someone with in depth knowledge of Quantum Theory who can think outside of the box of the science Orthodoxy read and evaluate the principles put forth in this paper. The science Orthodoxy, as the reader will see, has been at a standstill for a century, made errors – *I have even had to correct the calculation of the value pi, as an example, in this paper.* I have had to make so many corrections to the Orthodoxy, in fact, that the bulk of the paper is dedicated to such inverse thinking and correcting it. Once the corrections are made, and the proofs provided via inarguable theorems and axioms, the solid relationship between the Quantum Zeno Effect and General Relativity emerges, and the engine design, albeit inescapable to say, *becomes obvious and simple in design.*

The Rationale

Helium-3; nuclear fusion with no radioactive waste whatsoever. The product is pure proton, whose electromagnetic pulse is directly converted to electricity. It has already been done in the lab. Unfortunately, the isotope is so rare on Earth that it cannot be used as an energy source.

The Moon's surface is covered with it, in its regolith. Although fifty parts per billion doesn't sound like a lot, roughly a 5 by 5 mile patch of lunar soil cut 9 feet deep would power the US for a year with absolutely no radioactive or other pollutants. Patches of Earth this size are cut routinely for mining operations for various ores. There is enough Helium-3 on just the lunar surface to power Earth for thousands of years to come, eliminating the use of fossil fuels altogether; and yes, saving our planet from the seemingly inescapable 'Snowball Earth' extinction now imminent because of the burning of those fossil fuels.

The race to the Helium-3 is not just a race to the value in dollars, but a race to save our planet from the 7th Global Mass extinction now referred to as the Holocene Extinction. That is, the 7th Global Mass Extinction is not a possibility or a hypothesis, *we are in it; the **Holocene Extinction**.* The estimates vary, the average is 30,000 species go extinct each year. There are 9 million species on Earth. That means 300 years to a lifeless planet. All scientists agree that humans are the cause. [see 'Snowball Earth' Dr. William Joseph Bray, ISBN 978-1523613557]

At this time, building an infrastructure on the moon capable of harvesting this regolithic gold is impossible using chemical rockets. The engine design described in this paper bends space, not unlike our science fiction 'Star Trek' engines of the future. However, in this case, the effect is real, the energy consumption is negligible, and the Quantum Theory that describes it is quite tangible and real. Because the engine is gravitic, it 'doesn't care' how much mass it is lifting. A battleship loaded with bulldozers and building structures, and so on, doesn't matter. Faster Than Light is a secondary goal, lifting heavy loads into space is the immediate goal.

As an entrepreneur, you can comprehend 'owning' the Helium-3 on the lunar surface. I am not interested in money. That is the last thing on my mind. My intent is to prevent the 7th Global Mass Extinction, AKA, Snowball Earth, or at least, 'salvage' as much life of this planet as possible. That was why I was born. If this engine is not built, Snowball Earth (At least read the book and see the math and graphs and research) and the Holocene Extinction will leave not even a microbe on this freeze dried planet.

I am asking for you to pull in partners to afford me the research team I need to do the research and development. After that, *the engine is yours*. I do not want payoffs or royalties, just to build and test the engine. I estimate 10 years to the prototype.

This paper may seem to go off topic in many cases as a result, however, nothing could be further from correct. I have even had to correct misconceptions back to the calculation of the value pi. In some cases the work will seem to go off on a philosophical tangent. However, there are issues regarding Non-Locality vs. Superpositionality, Quantum Entanglement, and so on, which I address in a simple mathematical theorem that appears oddly philosophical in nature, as Von Neumann's approach to the Copenhagen Interpretation of Quantum Mechanics did at the time, albeit, explained here correctly for the first time, unabashed by mechanistic arguments.

This paper will describe how to produce an Alcubierre Space-time Manifold using the Quantum Zeno Effect to alter the progression of unitary time. The demands of General Relativity are clear that altering the progression of time must be accompanied by an alteration of the shape of space. The QZE is capable of both slowing and speeding the progression of unitary time. By careful manipulation of the QZE and Anti-QZE the Alcubierre Space-time Manifold is produced with very little energy other than that used for detection and measurement. No 'exotic' forms of energy are required, as proposed by the scientific orthodoxy.

CURRENT APPROACHES

This sub-topic is important to understand because financial, human, and other resources are being allocated toward examining these approaches that cannot yield a working engine. Furthermore, these current approaches make a mockery of the lack of insight of the Orthodox thinking in Quantum Theory as they propose absurd approaches to such a simple task.

My calculations regarding finance and other resources from proof of principle to building the working prototype of the engine in this proposal are significantly less than Capitol Hill spends on bottled water per year. The topics included in this section are actual projects that are budgeted and being researched at this time.

There has been no aggressive approach, theory, or agreed upon hypothesis to date, of how an alteration of the shape of space might be accomplished artificially. There are discussions regarding 'negative energy' but there is at this time no agreed upon definition for 'negative energy,' and it remains hotly debated, and is at the very least thus far unobtainable and elusive to produce, yield, or achieve.

There are a few experimental approaches under way testing methods how this might be accomplished, or at least otherwise defined. For instance, one effort involves passing bosons (photons) heavily stacked upon one another (LASER) passed through a pair of Casimir plates in order to 'evoke' a potential of negative quantum fluctuations (think of a balloon squeezed through a hole too small for it to fit through and upon emerging out from the between the plates there is a resulting 'gap' between the balloon's original girth and it's new lesser girth); passed through a toroidal magnetic field such that the virtual photons coupled to the magnetic field will form an interference pattern with this energy 'gap' (between the balloon's original and subsequent girth); and like a double slit experiment, focus on that portion of the interference pattern that has 'dark region.' The effort is actually secretive; however, from visual descriptions of the apparatus and what has been omitted it is not difficult to ascertain the entire setup and process. My description departs somewhat from the actual experimental setup I've anticipated. The addition of the Casimir 'filter' and the focus on the virtual photon interference patterns resulting in 'dark regions' in the resulting interference pattern are what I consider to be a correction to the experimental setup in order to derive a common definition of 'negative energy' in finite space-time. However, none of this is useful, nor shall it achieve anything. Virtual particles such as photons will not produce a vehicle. In fact, virtual photons are so common; they are responsible for the appearance of a magnetic field, for instance.

All of these measures are double talk used to convince an uneducated brass to cough up dollars for otherwise useless careers.

In addition, this discovery approach is far too young to be considered as a potential propulsion system at this time. In the Director's own words (who shall go nameless here), 'it might take two years, it might take two hundred years.' I call that 'fired.'

It is also important to note that this approach (mentioned above) states that it seeks a *negative mass budget*, in order to reduce the amount of thrust to propel an object. Modern definitions of *negative mass* are derived from Bondi, H. (July 1957). "Negative Mass in General Relativity". Rev. Mod. Phys. 29 (3):423. Bibcode: 1957RvMP...29..423B. doi:10.1103/RevModPhys.29.423

Before proceeding, I will use a simple example of Pythagorean's Theorem:

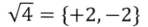

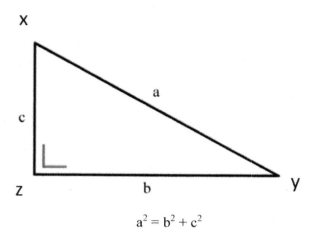

$$a^2 = b^2 + c^2$$

We solve for 'A':

$$(+,-)a = \sqrt{b^2 + c^2}$$

The value '-a' does not indicate or infer negative length, it is merely the difference in measuring the length 'a' from point x to y vs. point y to x, yielding the same value.

In similar fashion, it is non-sequitur to assume Einstein's originally published equation:

$$E^2 = m^2 c^4$$

He didn't actually simplify the equation to:

$$E = mc^2$$

In this case the correct solution would be:

$$(+,-)E = mc^2$$

Historically, the negative sign was simply dropped, as –E at the time was considered an artifact of the math, not part of the solution. However, later assumptions that negative energy, exotic mass via exotic matter as a result of arbitrarily moving the negative sign to solve for 'm', and so on evolved. Bondi and others have approached this much more exhaustively, but the principle is as simple as stated above. Negative energy, mass, and so on are indifferentiable from 'positive' energy, mass, and so on. Yet, there are those who work close to the financial resources who either believe this nonsensical approach and/or seek budget money (from those who cannot understand the arguments) to conduct pure research that can go on indefinitely with no reward. There is an Orthodox language for progress where there is lack of progress; much like 'I'm making progress by looking for my keys everywhere I have not yet found them.'

All values, including length, even when regarded as a 'static' measurement are time dependent. It is not possible to be one Planck unit of 'static' length without being one Planck unit of time distant. Space and time are indifferentiable, space-time. Yet, we still see hypothesis and arguments based on some number of dimensions of space and one dimension of time. If there are 11 dimensions of space and one of time, then there are 12 dimensions of space-time. There is no way around the argument without producing paradoxes, gaps in the description, and failures. Thus, we see entire theories (actually, hypotheses) laced with paradoxes and impossibilities that remain unresolved. Moreover, improvable by any means at this time, useless as a consideration.

Similarly, mass, energy, and all of the forces of nature are time dependent. There is no way to resolve time out of any system. Every equation regarding mass has a time element to account for. Mass means a 'particle' is present. A particle is a wave function (not a tiny cannon ball). No matter how close you zoom in, there is nothing actually 'solid' to be found. What makes 'stuff solid' is the Pauli Exclusion Principle, which is a way wave functions interact with one another. The principle is so powerful it keeps a Neutron Star from collapsing into a Black Hole. At just the point before such collapse, the Neutron Star is called 'degenerate,' it is one great big Neutron wave function; two down quarks and one up quark.

A wave function is time dependent. Every force of nature is a process and therefore time dependent. Every 'thing' observed and known to date are wave functions and forces, and as such are all time dependent.

There is no such thing as 'static energy potential.' It is a term taught in Physics 101 that carries through to adulthood, but does not truly exist. This possibility was eliminated by the Zero Point, which has a lower limit for any system based on Planck's constant, and therefore time dependent. Any value less than the Zero Point cannot exist in normal space-time, and therefore cannot transfer information in any form 'into' normal space-time by any other means other than a Planck Scale Traversable Einstein-Rosen Bridge. [1,2] The complexity of the temporal nature of the Einstein-Rosen Bridge is still not understood nearly a century later. However, it would require the spontaneous formation of these structures to transfer information from an encapsulated (sub Planck scale) domain to our Planck scale domain, assuming one side of the structure could be held to a sub Planck scale.

That is the basis for String Theory, rather, hypotheses, as a theory requires evidence. All String hypotheses in all of their iterations fail at the basic underlying beginnings of having a 'string' of 1 Planck length 'vibrate,' and then attempting to scale this up in some meaningful way:

1. Thorne, Kip S. (1994). *Black Holes and Time Warps*. W. W. Norton. pp. 494–496. ISBN 0-393-31276-3.
2. Ian H., Redmount; Wai-Mo Suen (1994). "Quantum Dynamics of Lorentzian Spacetime Foam". *Physical Review D*. **49** (10): 5199. arXiv:gr-qc/9309017. Bibcode:1994PhRvD..49.5199R. doi:10.1103/PhysRevD.49.5199

The most basic problem is in the wave function itself. If we take the most basic 2-dimensional representation of a sine wave of 1 Planck length and slice it up so that we can demonstrate small amounts of change occurring gradually along the length, 'x' we run into the problem that we are taking slices smaller than the Planck length will allow:

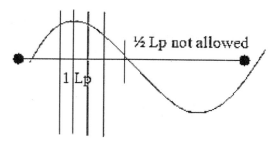

The same is true, of course, for the changes in 'y.' The original scaffold of String theory (hypothesis) was a string of one Planck length 'vibrating' in different modes. The only possible modes of action allowable in this case are {-1,0,+1} and no other values are possible. For a complete description of all of the possibilities of time allowable in this 2-dimensional holographic cosmos as a 4-dimensional façade [see Bray, Temporal Mechanics 101].

In addition, any unobservable or 'encapsulated' dimensions (such as proposed in various iterations of string theory) would each require a scaffold of Einstein-Rosen Bridges in order to manifest any properties into normal space-time. In this case, there has been no hypothesis regarding what form this *information* would take.

Since this is unlikely, we will forgo hypothesis regarding unobservable 'encapsulated' dimensions and work strictly with what is immediately observable, detectable, characterized, and 'real.' String and m-hypotheses which are untestable and unseeable will not provide an engine but will remain hypotheses indefinitely, possibly forgotten.

Part two of the argument requires that the Planck length is not the lower limit of space-time. This is a common argument put forth today by young physicists who write papers in various theories (which are actually hypotheses, theories require evidence, thus String Theory is actually String Hypothesis, and so on) which require slicing space-time into pieces smaller than the Planck length will allow. As a result, they simply disregard the Planck length as the lower limit in normal space-time without any evidence, mathematical argument, or any argument whatsoever. They need encapsulated dimensions, so they disregard the Planck length.

In order to evaluate this, we take a close look at the equations regarding the Planck length:

$$L_p = \sqrt{\frac{hG}{2\pi c^3}}$$

The equation is not contested (Wheeler). In the denominator we have pi and c, rock solid constants. In the numerator we have G, another rock solid figure. This leaves us with h, Planck's constant, the minimal amount of energy allowable in normal space time. The Planck Length, L_p is therefore completely dependent upon h. If, Then, Else: If L_p is NOT the lowest limit allowable in normal space-time, THEN h, *the quanta*, IS NOT the minimal allowable amount of energy allowable in normal space-time; ELSE, h, *the quanta*, IS the minimal allowable energy in normal space-time THEREFORE L_p is the minimal allowable slice of space allowable in normal space-time. We will see the Planck Units throughout this paper, so we need to clear this up now.

The 'Typhoid Mary' in every system is time, which cannot be eliminated under any circumstances. If one misconceives a 'Black Hole' as being a place where time has come to a stop, the equations actually state that time has *approach infinitely dilated, not zero.* In any case, the approach to the actual Swarzschild

Radius where time comes to a stop is asymptotic and never reached, but forever approached. Thus, in a certain sense, a true Black Hole cannot exist, but is always, forever, in the forming, but never achieved. In addition, since the Swarzschild Radius is the center of the phenomenon of infinite time dilation, a Black Hole can have no interior. That is, just as the material outside of a Swarzschild Radius is drawn to the radius, so is the material *inside of the radius* drawn toward the radius. Thus, the Black Hole becomes dimensionless in normal space-time, it only answers to a 2-dimensional solution.

There is no singularity, if it could ever be reached, the Swarzschild Radius is the ultimate end of a Black Hole. The backward thinking here is that because a thing is shrinking it will keep getting smaller until it reaches a point. However, the 'shrinkage' is toward the radius, not the center. The Schwarzschild radius *is* the singularity, albeit, not a zero dimensional point, but a 2-dimensional surface in 4-dimensional space-time. In simple terms, a Black Hole has no interior, it lacks the dimensionality.

As we shall see, the QZE, the manipulation of the progression of time by sustained observation drives every system, energy, mass, the forces of nature, and so on, and therefore 'paints' the framework of *everything that exists; wave functions and forces.* The QZE is *observation dependent, requiring an observer. Therefore, every system is observer dependent, or rather, observer interdependent.*

In these modern definitions for *negative mass* properties:

Inertial Mass and Gravitational Mass

The law for the gravitational attraction of a body of mass M for a body of mass m is given by

$$F = -GmM/r^2 = ma$$

Where r is the separation distance of the bodies and a is the acceleration of the body of mass m. Division of both sides of the equation by m gives the acceleration a of the particle as

$$a = F/m = -GM/r^2$$

This says that the acceleration of a body subject to gravitational forces is *independent* of its mass. A simple example of this is the Galilean observation that two objects of different mass accelerate toward the ground equally and later confirmed by the 'feather and hammer' demonstration on the lunar surface. This is a remarkable principle. If the inertial mass and gravitational mass are always equal then the motion of a particle of negative mass subject to the gravitational attraction of other particles would be the same as that of a particle of positive mass. Since the bubble model indicates that, the motion of a negative mass object is in the opposite direction from that of a positive mass object in a gravitational field it must be that the gravitational m_{grav} is the absolute value of the inertial mass. Thus the law of gravitation should be

$$F = -G|m_{inertial}||M_{inertial}|/r^2$$

Moreover, the result is that the alteration of the shape of local space-time will have no result on or otherwise result from the presence or application of *negative mass* that would be any different from the presence of 'positive mass.' This is extended to include all forms of mass-energy.

As one of the properties derived from Bondi's work, *negative mass* moves in the opposite direction when 'pushed.' This translates to an increase in thrust necessary to propel an object with an increased *negative mass budget* (which the government is funding because the brass cannot do the math). That is, by increasing a *negative mass budget* the actual thrust required to propel an object increases, not decreases. That is, since negative and positive mass are equivalent, increasing the negative mass budget, as it is termed, is akin to piling more conventional mass on our vessel to propel, in turn requiring more thrust, and so on. The people who fund these projects (control the money) do not understand this simple Equivalence Principle, and are wooed into funding snake oil sci-fi projects.

> *Time must never be thought of as pre-existing in any sense; it is a manufactured quantity.* --Hermann Bondi

Then there is the introduction of encapsulated unobservable dimensions, i.e., String Theory, m-Theory, and so on. Producing equations in an attempt to yield real results does not produce a real, but a speculative result, working with dimensions that have not been substantiated by observation or experimental evidence, and therefore not useful in producing a *real engine*. Any attempt to characterize these unobservable dimensions requires mechanisms capable of observing unobservable dimensions and quantitatively measure their properties. It is a palpable conclusion that this will not be achieved in the near future, if at all. Regardless of the 'truth' of the hypotheses, they will not produce a working engine.

Any engine design intended for real use in real time must be founded upon existing, validated, observable, and reproducible experimental data using models of physics that can be controlled, manipulated, and so on; not sci-fi hypotheses that cannot even be experimentally validated.

The QZE is founded upon such a framework to the extent that it has become an industry standard in approaches to quantum computing, now for sale. Moreover, the QZE is not regarded as a *phenomenon* but a race to produce the *best QZE* in order to beat the competitors in quantum computing in the market.

Another statement regards the idea that the symmetry of the Alcubierre space-time manifold 'does not know which way to go' because of its symmetry. The Alcubierre space-time manifold as originally stated by Alcubierre is not symmetric, but defies symmetry to the extent that physicists have misunderstood which way it faces and what the type and scope of its energy requirements are. The Alcubierre space-time manifold is by definition the most asymmetric 'thing' in the universe, assuming it can be manifest. The idea that it requires or produces a 'boost' is profoundly absurd and by definition of the manifold itself, denotes a complete misunderstanding of a space-time manifold of this sort. In no uncertain terms, if a 'scientist' came to me and said 1) the manifold is symmetric and 2) it requires a conventional chemical rocket inside the manifold so it knows which way to go, I would have him flogged for stupidity.

First, we can *dismiss* any notion of 'negative energy' requirements since the term 'negative energy' is not agreed upon nor is there any evidence in the hands of humans that demonstrate 'negative energy,' its existence, or its effects upon nature or anything in nature. We have absolutely no evidence of 'negative energy' appearing anywhere in our experiments or in nature and any references to 'negative energy' are purely speculative and unprecedented. As we explored, the best definition for negative energy is that it is indifferentiable from positive energy.

The statements regarding the idea that the symmetry of the Alcubierre space-time manifold 'does not know which way to go' because of its symmetry, and it therefore requires a 'boost' from a chemical engine source. This is a bizarre misconception that a chemical rocket placed in the center of such a space-time manifold will undergo a 'boost' in velocity merely as a result of being inside the manifold; dragging the manifold along with the chemical rocket, but at a higher velocity. There is no mathematical model or reasoning whatsoever that a chemical rocket within this space-time manifold will experience a 'boost' in velocity as a result of the space-time manifold described by Alcubierre. This is an argument used to impress men with clusters and brass on their chest, who require 'chemical rockets' in an explanation so as to feel they have a grasp of something they understand, but no knowledge of the subject, and fund the research. In fact, the idea that the thrust from an engine would make it past the spires of the manifold is of the utmost absurdity as they are by conventional definition contracted space-time.

It is the deliberate asymmetry of the Alcubierre space-time manifold, which provides velocity by altering the 'real' length, L_p, in front of and behind the engine, as will be shown later on in this text.

To date, every modification of the original manifold as described by Alcubierre that I have seen including 'rings,' manifolds within manifolds, and so on, produce no motion, no 'boost,' *nothing*. The original design is perfect, and requires no refinement. The only correction that I will make to Alcubierre's original manifold design will be to flip it upside-down and backward from its typical depiction, for reasons I will go into at length, then quantize it.

THE QZE AND TEMPORAL DYNAMICS

First, I want to approach the engine design in the first chapter that deals specifically with altering the shape of space-time by utilizing the Quantum Zeno Effect to shape at least one dimension of time, and leave General Relativity to demand a resulting reshaping of space. Space-time is thus reshaped quite precisely according to some guiding principles of the QZE and observed temporal change to produce an Alcubierre Space-time Manifold of exacting proportion to produce Faster-Than-Light travel without actual motion taking place.

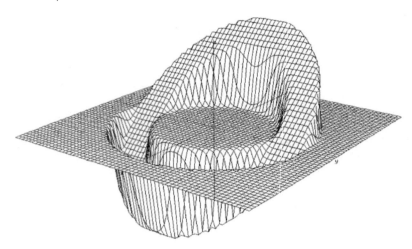

Everyone draws it and fantasizes about modifying it to perhaps require only a million universes worth of 'exotic energy' to manifest but the Orthodoxy fails to formally define such energy and therefore lays down yet another non-sequitur argument.

The background introduction material is extensive, but is necessary to clean up a lot of backward thinking in quantum theory and physics in general, so please read the entire paper carefully. The background material may seem obvious, but it is actually the presentation of century old equations that have been 'upside-down' and/or undefined and continue to disrupt common logic to this day.

Although Faster-Than-Light travel might seem like the stuff science fiction is made of, and beyond our immediate reach, this is simply not the case. Yes, it is impossible to go faster than light in an inertial frame of reference within the confines of this cosmos according to Special Relativity. However, in a system that is isolated from all of the above in General Relativity that is not a true statement. The Alcubierre Space-time Manifold as Alcubierre described it in 1994 is fine, just as described, except for being upside down and backward, and not quantized. The description regarding how it works is also upside-down and backwards. That is because the equations I will get into that are key equations in physics, a century old, have been upside-down for a century.

It actually works by stretching space to its target, then bringing up the rear. Contracting space is absurd. Our best example in nature requiring the mass of millions of stars contracting space does so with a factor a perhaps a factor of a few hundred to one. Stretching space can be infinite. The QZE has no upper boundary. The QZE acts directly on General Relativistic principles to stretch space towards its target without limit. There is no Special Relativistic consideration or limitations as the definition $c=L_p/t_p$ is never violated (where L_p is a Planck length and t_p is a Planck time interval).

An example is the photon. The photon does not contract space by consuming a million universes worth of 'exotic negative energy.' Yet, to the photon, the time and distance to every point in the cosmos is *zero*. It is thus Superpositioned throughout space-time. We see it as taking billions of years to get from A to B. But to the photon, the distance is zero. This is not a Faster than Light example, but as I will define later, represents backward (upside-down) thinking. The common argument is that space-time contracts for that one photon, billions of light-years of space-time contracting for one photon because Lorentz wrote his equation upside down because he misunderstood the outcome of the Michelson-Morley experiment. In Quantum Mechanics, the single photon superpositions, rather than the shape of the entire cosmos changing. The photon *stretches* from A to B and is at two places at once (AKA; superposition).

I will correct Lorentz equation later on and show the simple proof.

The technology to build the engine specified in this design exists now, would require approximately five years to achieve proof of principle, and another decade to scale the phenomenon up and construct the functioning prototype. The engine design and the working principle are in fact, obvious when explained. There is actually no 'new' technology going into the system, merely an application of existing phenomenon and technology observed in the laboratory in Quantum Mechanics for decades, treated mysteriously, unraveled into an engine design. The timelines involved are not 'guesses,' they are based on the state of current technology and estimated effort in altering such technology to produce the proof of principle and then the working prototype, given a working team of 12 physicists {2 mathematicians, 2 experts in phased array systems, 2 particle physicists, 2 computer scientists, 2 aerospace engineers, 2 physicists} working in relative isolation dedicated to the project full time.

It was in formulating Quantum Temporal Dynamics (by the author) that the dynamic behavior of multiple dimensions of time and their effect on space, matter, and the four known forces of nature became apparent, and discernable. It then became obvious to reveal the simple nature of the Quantum Zeno Effect such that it can be related to such primitive concepts as even Special Relativity, and of course, the demands of General Relativity, which provide the scaffold for reshaping space-time to produce a working Alcubierre Space-time Manifold exactly as Alcubierre presented in 1994. Interestingly, the Alcubierre Space-time Manifold under the conditions of the QZE requires extremely little, if any, direct energy to produce, mostly indirect energy in the form of computing speed and power. The approach of bending space-time by brute force is rather absurd, as it takes the mass-energy of an entire star to produce minimal curvature of space-time, as demonstrated directly in the 1919 solar eclipse that barely detected the bending of starlight around the sun. Nonetheless, the type of energy and the application of such energy to artificially bend space-time in the many ways presented to date by brute force remain mysterious because it is the wrong approach to take.

The demands of General Relativity, however, dictate that if the progression of time is altered there must be an accompanying reshaping of space, else violate General Relativity.

The simplicity is; the **Quantum Zeno Effect** is the suppression of 'unitary' time by constant observation:

> 'The Quantum Zeno effect is the suppression of unitary time evolution caused by quantum Decoherence in quantum systems provided by a variety of sources: measurement, interactions with the environment, stochastic fields, and so on.' [T. Nakanishi, K. Yamane, and M. Kitano: Absorption-free optical control of spin systems: the quantum Zeno effect in optical pumping Phys. Rev. A 65, 013404 (2001)].

Where the term 'unitary' refers specifically to the progression of one Planck interval of time to the next, that is, a unit of time. Physics has no explanation how one Planck interval of time (approximately 10^{-44} seconds) advances to the next, referred to as the 'Planck Flow.' Time itself is not continuous, but flickers like an old style movie projector, yet too fine to detect by even our most advanced instrumentation. This is where the term 'Zeno' comes into the term Quantum Zeno Effect: [See Bray ISBN-13: 978-

1478176435 2001] We call this the Planck Flow, the seeming continuity of space-time. To date, there is no explanation for it, nor has there been a notable attempt at it. I will explain it later in this paper.

Let me just make it clear that the founders of Quantum Theory had the Planck interval written in stone. Today, as a result of hypotheses threatened by the presence of the Planck interval we see certainly much less of such a dependency discussed, but only in hypotheses. However, the infinite divisibility of space-time requires infinite energy from the QED vacuum energy to explain the presence of just one cubic centimeter of space. By dividing that same cubic centimeter of space into Planck intervals we avoid infinities, and the same holds true with all space-time. All space-time, even infinitesimals, require producing infinite amounts of energy to sustain for infinitesimal amounts of time if space-time is infinitely divisible. This is because of the simple axiom, any infinite thing divided by n equals infinity:

$$\lim_{x \to \infty} \frac{x}{n} = \infty$$

In this case, even the smallest space-time interval requires infinite energy to produce. Quantizing space-time eliminates infinities.

> Zeno was a philosopher of Plato's time, and his ancient question, 'if you take an infinitesimal slice of time of an arrow in flight (an *instantaneous* snapshot), it is not moving, how then is it moving?' Zeno was asking questions that directly indicated an understanding of infinitesimals circa 500 B.C., thousands of years ahead of the discovery of infinitesimals. Today the lower limit of this phenomenon is not an infinitesimal but Planck's constant, quantized space-time, 10^{-44} seconds, 10^{-35} meters.
>
> The argument that you add up an infinite number of snapshots each taken at an infinitesimal slice of time is an error of calculus that does not apply here, historically referred to as a Riemann sum. The problem in Zeno's argument is that if it is not moving in one frame (snapshot) how can it get to the next frame (snapshot)? You would be adding up an infinite number of snapshots of the arrow in the same position. That is, no Planck Flow.
>
> This is the same exact problem we run into in every attempt to explain the 'Planck Flow.' There is no mechanism in Quantum Mechanics to explain the continuity of time, nor of space.
>
> To make matters worse, Holography has even eliminated time altogether from the dimensionality of the universe, reducing the cosmos to a 2-dimensional surface (I refer to as a Schwarzschild surface, explained in another text).

In 1977 George Sudarshan and Baidyanath Misra of the University of Texas discovered that **if you continually observe an unstable particle it will never decay.**

- Sudarshan, E.C.G.; Misra, B. (1977). "The Zeno's paradox in quantum theory". *Journal of Mathematical Physics 18 (4): 756–763.*

Heisenberg's Uncertainty Principle stated that as you take a smaller and smaller slice of time, the uncertainty of the particle's position, momentum, and energy increases? Well, in this case, Sudarshan and Misra discovered that as you take those *measurements* at smaller and smaller intervals, slices of time, bordering on continuous observation – if you never take your eyes off it, *nothing changes; it becomes frozen in time, or slowed to a near stop.* We look at the HUP with respect to position and momentum:

$$\Delta x \Delta p = \frac{h}{2\pi \Delta t}$$

Here we can intuitively see that if Δt is as small as it can possibly be, $1t_p$, the HUP at least can account for the idea that we cannot state the decay particle as having progressed from the confines of the nucleus, or the orbital, etc. This is akin to the idea that Zeno's arrow represents infinite uncertainty of where it will strike because with only one snapshot, we cannot even say whether or not it is falling straight down. That is, a very large momentum and position uncertainty means we have no idea where the thin is, in which case we have no idea where it is going.

If on the other hand Δt is infinite in this equation, we can then liken the infinitesimal momentum and position uncertainty as a snapshot of uncertainty. That is, we can say that we are certain to an absolute degree what its momentum and position are, but as a snapshot value, not what its next position will be, regardless of momentum (vector information). The reason is that the vector information is an infinitesimal, inadequate to formulate a complete picture of the arrow's next position, even on a Planck scale, which is much greater than an infinitesimal.

With respect to energy, we have to look at that snapshot from an entirely different perspective. Think of a snapshot as one infinitely long stare, which is what the QZE is:

In this case, the Heisenberg Δt is infinitely large, not small, rendering ΔE as zero.

$$\Delta E = \frac{h}{2\pi \Delta t}$$

Such will be the case as we unravel Schrodinger's equations and the like. They will all behave characteristically as time dependent (they are all time dependent) and answer to the Sudershan Misra statement regarding this approach to Δt. In one sense, it is one Planck interval of time, and another sense, an infinitely long stare, like a photograph.

As we raise the observation rate slightly above t_p (in integer multiples of t_p only), extremely rapid observations, the uncertainty of where the arrow will be in the next frame is high enough that over enough such rapid observations it appears that the arrow's flight has slowed down. That is, the more rapid the observations, the greater the *Uncertainty* of the arrow's next state in the next frame. This is the QZE, and to date, all proposed mechanisms attempting to explain it are gibberish. Eventually, your observations become so rapid (continuous) that the *Uncertainty* of the arrow's next state in the next frame forbids its progression altogether and the arrow ceases to progress altogether. During this dive into increasingly more rapid observations, the growing *Uncertainty* of the arrows next state in the next frame is causing it to stall in its progression, causing the arrow to slow down. In terms of QZE, this is where we are staring at a photograph of the arrow frozen in flight, it could be flying forward in the photo, or falling in a dead drop toward the ground, being beamed up by Scotty, or being tossed straight up into the air; complete Uncertainty.

Radioactive decay is the gold standard by which we measure and calibrate time on all levels, from the accuracy of a trillionth of a second by cesium observation (US Naval Bureau of Standards) to the billions of years marked by uranium/helium isotope counting. Radioactive decay, aside from the QZE, is immutable, so immutable that it was used to measure the effects of Special Relativity.

The Quantum Zeno Effect actually causes the rate of radioactive decay to change. It is a temporal marker, telling us that the rate of progression of linear time has been artificially altered. It is not a trick or sleight of hand, and there is no workaround argument as proposed by many a blogger or would be riddle solver. I have read countless papers by individuals who claim to have 'solved the QZE enigma' with mathematical gibberish by using models of radioactive decay and a complete miscomprehension of time and 'particles' as wave functions rather than tiny cannon balls.

The worse I have seen to date, hailed by computer geeks educated in basic logic but not in advanced mathematics, was by a cryptographer, Turing, referred to as the Turing solution, essentially is a complete restatement of Zeno's infinitesimal, not a solution. But since computer geeks like it, it is a solution gone viral, albeit, gibberish.

The QZE does what no other phenomenon other than Relativistic time dilation can do; it slows the progression of time. In this we find that the QZE, just as in Special and General Relativity slows the actual progression of real time. (In GR we simply sink the radioactive source deep within an intense gravity well). The QZE does this without speed (Special Relativity) or gravitation (General Relativity). Further on I derive the math, not unlike relativistic properties, that characterizes these systems. That is, there is a third system that alters the progression of time, and as such, General Relativity does not take a siesta from the laws of physics, GR demands that space reshape itself according to all of the common principles of General Relativity. Likewise, this is where we build the scaffold of our space-time manifold. However, this third system does not require relativistic speed or extreme mass conditions, only detection and measurement (observation). The QZE is fixed in space, void of velocity or inertial frame of reference, and any massive object. By detection and measurement only, produces the same time dilation *and antidilation effects* as Special and General Relativity. In SR and GR we do not see actual antidilation temporal effects, we see relative antidilation temporal effects. That is, an object deep within an intense gravity well will perceive a temporal antidilation effect of an object far from the gravity well. But in the case of the QZE, we measure the target as dilated or antidilated, directly regardless of its environment.

In one sense, we place an observer in a stationary frame of reference measuring the length of a photon emitted from a source speeding **away** at a velocity that results in a wavelength twice its value (a red-shift factor of 2). This, incidentally, was how Einstein originally thought relativistitics up while riding the bus to work.

The observer measures the photon as viewed head-on (yet moving away) by timing the instance of the wave from start to finish. As observed head-on (yet moving away), the stationary observer observes that photon for twice as many Planck intervals of time than if the object emitting the photon were stationary ($2n\ t_p$), as measured by his own watch. Twice as many Planck intervals of time have elapsed in order to measure the same photon as seen moving away. We can argue that the passage of the photon from start to finish has taken a longer path.

Now, we know that it is red-shifted 2:1, *but where is the actual space that the additional length is occupying coming from?* By my watch, the wave is $\Delta t = 2n$. I am measuring at a faster rate by a factor of $2n$. As an example, we look into deep space and see the most distant galaxies at the visible horizon at 13.8 billion light-years away and say the additional space is due to the comoving distance, which is actually now about 45 billion light-years away, a longer path of *real distance.*

In my little experiment, a red-shift of 2 leaves a dilemma of where the extra space is coming from. The academic (wrong) argument that it is merely an observed and not a 'real' effect is gibberish. The equations are the same for time dilation, mass increase, and so on, all 'real' measured phenomenon. The additional space must be that somehow my light is taking a longer path, but how?

What seems like a straight line in a 2-dimensional representation of a 4-dimensional façade becomes questionable at best. In every equation I have seen to date translating this matrix I do not think anyone can describe the end product, which we know is real, nor the commuting to that state. Leonard Susskind at best had to work in 5-dimensions to feel comfortable with the outcome. Frankly, our entire view of what is straight is a non-sequitur at this point. Just as M-theory expanded our world out to many dimensions, crushing everything we can perceive with our own eyes and senses down to two dimensions makes everything in our sensory bubble a non-sequitur. Therefore, it is not amazing to me when I consider a beam of light traveling in a straight line taking a longer path when traveling in what looks to me to be the same straight line, but taking longer to do so. That is, we are still asking the question, how many dimensions are there? And, where is the extra space coming from?

If we regard the passing photon as viewed head-on (yet moving away from us) as a time progressive event, and furthermore regard the length and duration of the photon as fixed and immutable (the speed of light constant in all frames of reference), such as we regard the progression of radioactive decay as both fixed and immutable (provided it is both stationary and in our stationary frame of reference), we are forced to conclude that we have taken more rapid measurements of the photon from the perspective of the photon ($2n\ t_p$ within our fixed time reference). That is, we have taken twice as many measurements for that photon, as viewed head-on (as the photon speeds away from us), more rapid measurements of the same photon.

From the observer's perspective, the 'real' length of the photon has increased. Thus, the path from the head of the photon to the rear of the photon has increased. *More rapid observation, more sustained observation, longer path, space has altered shape.* However, the passage of the photon, keeping the velocity of light as a constant in all frames of reference, has crossed the same distance in a greater number of Planck intervals of time as measured by the observer. It can be said that the passage of the photon has taken a longer path from start to finish, or that we have taken more rapid measurements of the same photon. However, from the photon's perspective, since more Planck intervals of time have been taken to measure its passage from start to finish; the photon does not observe its own length as having changed. From the photon's perspective, according to Special Relativity (if we assume for a moment that Special Relativity applies to a mass-less photon) the observer has 'sped-up.' Actually, the photon perceives the entire cosmos as an infinitesimal.

This is a twist on the 'twin paradox' where the traveler speeding away from the stationary observer sees the passage of time for the stationary observer as having increased (by the end, return portion of the total trip; but not necessarily at each leg of the trip). Whereas most incorrect descriptions of the Twin Paradox refer to the acceleration of the ship at its turning point, the 'paradox' effect is caused by Lorentzian changes, dilation and anti-dilation.

In this case, however, we are all considering that the passage of the photon, according to the stationary observer, has taken a longer path from start to finish of the length of the photon (dilation). If one argues that, the photon has merely increased in length (dilation) that is exactly the point. The path of the photon from start to finish has taken a longer path across the head of the stationary observer.

In a little bit, I will show that the conclusion is that the value, L_p has changed, and derive the equations that conclude such.

This is important, because construction of Alcubierre's space-time manifold requires curving space-time intensely inward on one edge and intensely outward on the opposite edge. And the application of brute force and energy cannot be controlled in thus fashion. However, the detection rate of a phenomenon circumventing the 'engine' can be done as precisely and rapidly as the limit of our technology will allow.

There is a difference then between observation and constant observation. The obvious one being that common observation comes with what is referred to as flicker rate in animals. Flicker rate is equivalent to saying acquisition rate. The human eye or senses does not have infinite capacity for data acquisition. In human vision, for instance, different tests have placed that flicker rate somewhere around 200 Hz. The QZE is a mechanized process within the nanosecond range, well beyond Von Neumann's Wave Function Collapse window.

What does it mean? It means that if you stare at the snapshot of Zeno's arrow frozen in flight forever, it will never make it to the next frame. No human can do this, only a machine. However, when you as much as blink (metaphorically), and it is suddenly in the next frame. If a machine can keep a continuous eye open on the phenomenon that phenomenon will not change state; no particle will decay, virtual particles will cease to flood in from the QED vacuum, and so on.

In addition, you cannot possibly observe it in the act of making it (the arrow) from one frame to the next, because that would require watching it continually, and if you do so, it will never change to the next frame. You will never, ever, be able to catch that arrow in the act of going from one frame to the next. There is no actual passage from one frame to the next, time is not 'smooth' but quantized, it must move in the form of a quantum jump. (This 'jump is also defined in the first chapter of the second half of the text [not included in this paper] describing motion on a quantum scale as only occurring at the speed of light, with no other possible speed). Furthermore, *the arrow knows you are watching, and will quite intentionally wait until you stop looking to jump to the next frame; that is what decades of experimentation indicate, evidence of Von Neumann's approach.* As esoteric as that sounds, it is an accepted formality in the description. However, this agrees with those of Von Neumann's Copenhagen Interpretation of Quantum Mechanics, which is a cornerstone in the philosophy of Quantum Theory for over half of a century. One key is that quantum jump from one Planck interval of 10^{-44} seconds to the next Planck time interval of 10^{-44} seconds. Time in this cosmos is quantized. There is no workaround for it.

Schrodinger's Cat – the short version

The response to that last paragraph goes back to a thought experiment dating a century ago which to this day remains unresolved. The 'Cat Paradox,' as it has become known over the past century is an entire later chapter [in another text]. The key is in Superpositionality and Quantum Entanglement. Superpositionality and Quantum Entanglement are important concepts in Faster Than Light technologies because there is no common 'now' that exists between any two points in space-time anywhere in the cosmos, all the way down to one Planck interval removed. Now we are taking two points in space-time, perhaps light-years apart and defying that basic principle. At the end of the day, we will have defied a lot of principles to get from A to B, not the least of which is the redefinition of what constitutes 'now.' In fact, according to the way I have designed this engine, if the trip is one light-year, the nose and the stern will occupy a common 'now' which are one light-year apart for a brief moment, or something to that effect.

I'll very briefly describe it here but save the details for a detailed chapter:

> A detector is placed inside of a box to measure the exact moment when the carbon-14 atom in question decays. A cat is placed in the box with it. At the exact moment the carbon-14 atom in question decays, a hammer is triggered that falls and breaks a bottle of poison that kills the cat. The implication is that the macroscopic condition of the cat, being alive or dead, is linked to the Quantum Scale event of the carbon-14 atom's decay. Thus, the state of the cat, dead or alive, exists as a Superposition of being both dead and alive until such time that the box is opened and the observer identifies the state of the cat.

*We must define **consciousness** in terms suitable for Quantum Theory in order to comprehend the Quantum Zeno Effect.*

The issue and resolution to the cat paradox is, simply put, that no one to date has assigned consciousness to the cat, at which point the entire foundation for the 'paradox' fails to exist in the first place. The paradox is a thought experiment to resolve the issue of Superposition as the result of consciousness, in this case by assigning an impossible superposition state, alive and dead, to a living thing, and arbitrarily negate any thought of the cat as being a conscious life form. The cat being conscious places the burden of consciousness in the Von Neumann Copenhagen Interpretation of Quantum Mechanics (the conscious observer causes Wave Function Collapse) upon the cat, negating the role of the secondary, now passive observer, Schrodinger. The debate regarding the Von Neumann approach that humans collapse wave functions is troubled enough, now we attribute such characteristics to cats. However, in Quantum Mechanics, there is no magic line that defines where the cutoff for such Wave Function Collapse begins or ends (with respect to intellect or species), and most importantly, *in Quantum Mechanics, philosophy, medicine, all science and religion,* **there is no definition for consciousness altogether**, *nor are there definitions for life or death within the formalities of quantum mechanics. Even if we had a clinical definition for such, and we do not, that definition does not extend to quantum mechanics. The closest definition we have is a legal definition for death* [**William J Bray Quantum Physics, Near Death Experiences, Eternal Consciousness, Religion, and the Human Soul Edition II, 1988, Legal Definition for Death**] *We have an experiment dependent on three values none of which are formally defined, making the entire experiment non-sequitur.* [For a detailed approach to a definition for consciousness suitable within the framework of Quantum Theory, philosophy, and religion, see Bray, Quantum Physics, Near Death Experiences, Eternal Consciousness, Religion, and the Human Soul Edition II]

I do not know how many non-living things can be regarded as conscious in this cosmos. However, this 19th century thinking that a living thing is not conscious, sparking a ludicrous debate regarding the superposition state of a cat that to this day is still regarded as non-conscious is – *a spectacle of human absurdity. The Gedankenexperiment seems to lack Gedanken.*

The 'Cat Paradox' is an issue debated by the greatest minds of the founders of Quantum Physics. The implication is that the state of a, in this case, radioactive isotope (to replace the conscious cat) is Superpositioned in both a decayed and non-decayed state until an observer observes the state of the isotope. Variations on this cat 'paradox' are numerous. In each case, every variation and scenario debated, the outcome is always a Superpositioned state, never a mechanistic outcome. No one has ever found a non-Superpositioned, mechanistic outcome to this paradigm. Of course, every revision I have seen to date has had a live/dead yet not conscious cat, which in itself cannot be defined without a definition for consciousness, life, and death. That is, a live cat that lacks consciousness. A dead cat that does not fulfill the legal definitions for death I have so painstakingly taken the time to use my paralegal skills to research and record in my text. Interestingly, all of the legal definitions seem to pass on the definition to the discretion of the attending physician(s) with the exception of one statement: *irreversible*. However, without a formal definition for life, that term *irreversible* is also non-sequitur; *irreversible life*. In simplest terms, you can kill me, I go to Heaven, that is *irreversible*, but it is not *death*. I can reincarnate, which then qualifies it as reversible, and also not death. The law has to incorporate the various religions and belief systems of man in the USA. You begin to see the complexity of the problem not only in esoteric form of a belief system but in reality we have no meter stick that can measure the temperature in Heaven.

We are NOT off topic. We are attempting to define *observation*. Where we lack a definition for consciousness, we also lack a definition for consciousness. The question is, does 'observation' specifically refer to *conscious observation*, as does the argument in the cat paradox, the double-slit experiment, and so on?

Now let's really mess with the Cat Paradox. *Let's have Schrodinger and his box speeding away from us at relativistic velocity, such that it approaches infinity.* The state of the cat, or actually the isotope (replacing the cat or any temporal phenomenon), is dependent on Schrodinger's observation of its state. Until Schrodinger opens that box, the isotope remains Superpositioned, frozen in time, in multiple states simultaneously according to the stationary observer. Therefore, go ahead Schrodinger and open that box. However, you are an observer watching Schrodinger preparing to open the box while he is moving at relativistic velocity with respect to you.

As with all descriptions of relativistic time effects, Schrodinger is unaware of any change in the progression of linear time. He will open the box in his local real time, let's have him do it right now, and now he is observing the isotope, meaning that it is no longer Superpositioned in Schrodinger's frame of reference (Special Relativity). *However,* you see Schrodinger apparently frozen in time because he is moving away from you at relativistic velocity, he has not opened the box in your stationary and distant frame of reference, and therefore the isotope remains Superpositioned in multiple states with respect to you. *Distant* is an important factor to enter here, because just as Bell's Inequalities imply, any information regarding the state of whatever is in the box as Schrodinger opens and observes it cannot become a reality to you at any velocity faster than light, such that if the distance is one light-year it will be a year before that superposition becomes a fixed result. Thus, by having Schrodinger speed away from us with his box at relativistic velocity we have indeed added a new layer of complexity to the paradox, or so we call it, provided we do not use a living subject as the subject. Second, the Inequality only becomes evident if and when Schrodinger slows to non-relativistic velocity, other he remains frozen in time indefinitely to the stationary observer.

Thus, relativistic velocity in this scenario, speeding away from us, results in continuous observation by you, (black-shifted) absolute Uncertainty, chaos. It is only when the speeding ship carrying Schrodinger slows down that the $f_{observed}$ in the equation becomes less than infinity, observation is not continuous, *the arrow makes it to the next frame,* Certainty increases, and Wave Function Collapse, *order* occurs.

Two observers. One observes a non-Superpositioned outcome for the isotope (Schrodinger), the other (you) observes a Superpositioned state for the exact same particle of isotope. So now we have our particle of isotope existing in a completely new and novel dual state of being both Superpositioned and non-Superpositioned simultaneously. There is no name for this bizarre dual state. To the best of my knowledge, it has never been described before, because everyone cannot get past the cat.

The odd thing is, every particle in the cosmos exists in this state. Rovelli has gone to 20 years of trying to explain and/or define 'Relational Quantum Mechanics.' The idea is that 'what if two observers get into a unique situation where they do not observe a phenomenon in the same state?' He refers to this novel condition as observer dependent. Unfortunately for Rovelli, the obvious condition in space-time is that it is not possible for any two observers anywhere in the cosmos, even if they are just one Planck Length apart, to observe a phenomenon in the same state.

Furthermore, we do not actually need relativistic velocities to create this dual-Superpositioned state. The same holds true for observation of Schrodinger from any distance, all the way down to 10^{-35} (one Planck length) meters away. As we will examine in detail, there are only two possible velocities to travel one Planck length, at $v=c$ and $v=0$ (described later). Thus, the Superpositioned state of the particle of isotope remains Superpositioned for each *potential conscious observer* throughout the cosmos, even if its state is known by one observer. Arguments to date regarding the state of a system have remained highly localized, negating any universal content or scenario. The assumption has been that if the particle's state is known by a single observer that the state of the particle is then somehow absolute. There is, however, nothing to suggest this is the case. Quite the contrary, we have a reasonable argument otherwise.

We are all familiar with the Bob and Alice scenarios. Once Bob knows the quantum state of particle A, Alice *instantaneously* knows the state of particle B, and so on. However, as sequitur as this argument sounds, in real space-time such *information* cannot be communicated from Bob to Alice or visa-versa regarding confirmation of such faster than light, and thus the argument is non-sequitur. There is no common 'now' between any two points anywhere in the cosmos, even one Planck length apart.

The argument of observing quantum entanglement, for example, in a bubble chamber from a position above the plate (looking down at a photographic emulsion) is an error in dimensionality. Two signals reach a detector simultaneously does not suggest that the signals exist simultaneously or were even sent simultaneously or were equidistant from the detector. As the QZE suggests, we can literally freeze Bob's signal merely by looking for it.

There will be a number of Bob and Alice, Bell's Inequality rebuttals. However, they remain a 'paradox' to the rebutter's, and rebutting with a 'paradox' is a non-sequitur. I will dismiss Bob and Alice a bit later on.

In general, we have a cosmos, a closed system, treated like a box, not as though it were contained within another system; although our physics at this point (in particular m-theory and so on) require the cosmos to be contained within a larger, if not infinite system. Bob and Alice are always contained within the system and never appear, see, or exchange information outside of the system; *outside of the box,* and a 'paradox' ensues. Smoke a joint and befuddle one's own mind with infinite non-sequitur riddles and count oneself among the intellectual virtuoso.

The simple solution is that if the 'box' (our cosmos) Bob and Alice are contained within is finite, and our cosmos is contained within an infinite system, then:

$$\lim_{x \to \infty} \frac{n}{x} = 0$$

The box shrinks away to an infinitesimal, where *n* represents our cosmos, and *x* represents an infinite system, regardless of proximity, since by definition, an infinite system would make all other systems subsets.

Definition: (Definition of a limit at infinity)

We call L the **limit of** $f(x)$ **as** x **approaches infinity** if $f(x)$ becomes **arbitrarily close** to L whenever x is **sufficiently large**.

When this holds we write

$$\lim_{x \to +\infty} f(x) = L$$

or

$$f(x) \to L \quad \text{as} \quad x \to \infty$$

Similarly, we call L the **limit of** $f(x)$ **as** x **approaches negative infinity** if $f(x)$ becomes **arbitrarily close** to L whenever x is **sufficiently negative**.

When this holds we write

$$\lim_{x \to -\infty} f(x) = L$$

or

$$f(x) \to L \quad \text{as} \quad x \to -\infty$$

So, in this case, we write:

$$\lim_{x \to \infty} \frac{1}{x} = 0$$

and say "The limit, as x approaches infinity, equals 0 ," or "as x approaches infinity, the function approaches 0 ".

We can also write:

$$\lim_{x \to -\infty} \frac{1}{x} = 0$$

because making x very negative also forces $\frac{1}{x}$ to be close to 0 .

$$-\infty = +\infty$$

As for the cosmos it has a well-defined lower limit of the Big Bang and a well-defined upper limit of the present. By definition, having a lower limit defines a thing as non-infinite, it has a lower boundary:

$$(-\infty = +\infty)$$

The upper limit is the present, and not a microsecond beyond. The temporal limits for the cosmos are thus bound on both sides and finite. Any hypotheses otherwise are just guesswork as to the fate of the cosmos, and the source it completely unknown.

This allows Bob and Alice to be both inside and outside of the system, the box, simultaneously. The reason this bit of philosophical prose is here is because there is more to the Quantum Zeno Effect than equations. Some of the lighter material is noted above. The fundamental element of the QZE is in the Von Neumann Copenhagen Interpretation of Quantum Mechanics.

Why can Bob and Alice perceive both inside and outside of the box? According to the Von Neumann Copenhagen Interpretation of Quantum Mechanics, *observation* specifically refers to *conscious observation*. And for lack of a suitable definition within the framework of Quantum Theory, I have to assign *consciousness* as an infinite thing residing within that infinite domain, else, inert as any particle of space dust, with no suitable rationale for rendering it as not inert.

The fundamental working principles of the working engine are not purely mechanistic. As you can see from just a few of the working paradoxes involved in the nature of the machine.

This is not going off topic, nor is it stringing into a philosophical argument. The QZE was designed by nature to be the mechanism that makes one Planck interval of time 'tick' to the next, provide the natural flow and order of the seeming continuity of time on our macroscopic scale. On a Planck scale, there is no known mechanism (other than what I will describe here in this paper) for the continuity of one Planck interval of time proceeding to the next. In Holographic Theory, time should not exist at all.

As for how much or to what degree this slowing of the progression of unitary time has been achieved in the laboratory, the answer is a complete stop (t_0 -> t_∞). In terms of gravitation, this is equivalent to a Swarzschild Radius of a Black Hole (non-rotating) where time dilation is infinite, mass is infinite, and space curvature is infinite. Thus, any degree of space-time curvature can be achieved on a local scale via the

QZE. To reiterate, any alteration in the progression of time must be accompanied by an alteration in the shape of space or violate General Relativity.

The fact is, the QZE has surpassed the Black Hole by far with respect to time dilation.

$$t' = t_0 / \sqrt{1 - \frac{2GM}{rc^2}}$$

Note here that gravimetric time dilation is usually presented upside-down, with t' getting smaller. This is inconsistent with t' in Special Relativity, and also nonsensical, given that in a gravity well, if a clock is ticking at ½ the rate, t' has increased by a factor of 2, not decreased by ½. I therefore write gravimetric time dilation in this form for consistency and logic sake.

The equation for time dilation at the Swarzschild Radius of a black hole is asymptotic. That is, a black hole never actually forms because the approach to the Swarzschild Radius is asymptotic. A black hole is forever in the making. The QZE, however, is capable of bringing time to a complete stop. A black hole is incapable of this. This is an important distinction, because as we run into the 'top hat' function a bit later on, making that 90 degree bend in light would not be possible by any other means.

Producing the QZE is the first proposal for the artificial reshaping of the curvature of space, and requires no direct and certainly no exotic forms of energy. Although the QZE has been produced in the laboratory countless times in the laboratory and in the field before, the mechanism here must be scaled up in size and intensified in a very specific form in order to produce an exacting Alcubierre Space-time Manifold in order to yield the desired effect of traveling without motion at any velocity. We'll examine the exact details as we proceed. It should be noted that although the QZE can be produced in the laboratory, no one knows why or how it works, no one understands it, and it has only been done on microscopic scales of a few atomic radii.

Understanding the extensive background discussion is imperative, so please be patient as we do so.

There is no lacking of definitions for the Quantum Zeno Effect, most applying a hypothetical cause as to the nature of the phenomenon. For example, we look at what is referred to as Turing's 'Paradox,' which is commonly stated and repeated, but obviously not understood:

> It is easy to show using standard theory that if a system starts in an eigenstate of some observable, and measurements are made of that observable N times a second, then, even if the state is not a stationary one, the probability that the system will be in the same state after, say, one second, tends to one as N tends to infinity; that is, that continual observations will prevent motion …
>
> — Alan Turing as quoted by A. Hodges in Alan Turing: Life and Legacy of a Great Thinker p. 54

In this case, if we look at Zeno's arrow in flight, negating for a moment that space-time cannot be infinitely divided (limited by the Planck interval), and divide an observation of the arrow in flight into an infinite number of slices, what we have done is rendered it into such a state that we have to pour through an infinite number of observations in order to witness an infinitesimal movement of the arrow:

$$\lim_{x \to \infty} \frac{n}{x} = 0$$

That is, Turing's 'Paradox' has merely sliced an observation into an infinite number of slices, meaning that we have to look at each of an infinite number of slices to witness an infinitesimal change. This is not a 'paradox,' merely the result of infinitesimal calculus. Turing was a cryptographer, not a physicist. He essentially repeated Zeno's Paradox 2500 years later word for word, the difference being that Zeno of Elea was speaking of infinitesimals more than 2000 years before they would be characterized by mathematics. Nevertheless, an infinitesimal cannot exist in normal space-time; therefore, Turing's Paradox does not apply to the QZE in any case. The trend Turing speaks of does not apply in a quantized system, at least until you begin reaching slices in size on the order of that quantization. The QZE is apparent at many orders of magnitude larger than this, temporally. The reason the trend does not exist in a quantized system is that the limit 'x approaches infinity' is replaced by 'x approaches tp,' and the statement 'x approaches tp for f(x) =0' is not true.

$$\lim_{x \to tp} \frac{n}{t_p} \neq 0$$

Turing's paradox although mistakenly equated with the QZE, is a null statement in every scenario. It is simply an incorrect statement.

The reason I bring this up is that Turing's 'Paradox,' if anyone actually thought about it clearly, supposes the QZE is a phenomenon of preferential perspective, as suggested in the first equation. That is, slicing an observation into an infinite number of slices:

$$\frac{n}{\infty}$$

Requires an infinite number of observations to witness an infinitesimal change:

$$\lim_{x \to \infty} f\left(\frac{n}{\infty}\right) = 0$$

There are a host of similar attempts to explain the QZE that are accepted and given the reverence of a 'paradox' where none exists, they are simply wrong, and as I've observed, propelled by the young or people not in their field of expertise.

The QZE is not a phenomenon of preferential perspective. Several industrial processes already employ the QZE as a fundamental, quantum computing being at the forefront. Qubits are actually kept alive long enough to perform calculations via the QZE. Without the QZE, quantum computing would be hopeless. Magnetometers also use the QZE, along with several other industrial gadgets and processes. The QZE has become an industrial part of the 21st century.

Both the suppression and acceleration of unitary time have been achieved at will in the laboratory in uncountable cases over the decades. The simple demands of General Relativity require that an alteration in the progression of time be accompanied by a reshaping of local space. This reshaping of space is the means by which an Alcubierre Space-time Manifold is produced, rather than by brute force, or by detection and measurement at very high speeds of a large population of quantum scaled events surrounding the engine. However, we will never know this under the conditions in which we employ the QZE today, under microscopic if not subatomic scales, only. The brute force technique has obviously had no rational proposal to date.

Scaling up the QZE is the proposal of this paper.

In this paper, I have chosen the particle spin characteristics of electron-positron pairs because they can be produced in vast numbers (high luminosity), do not suffer the 'measurement problem,' (a common issue where the observer gets confused regarding the observation and interfering with the observation, producing the result) travel at high velocities, and be directly measured with great ease at high speed via a Stern-Gerlach approach. The details of the engine emitter and detector design, including the mathematics of the measurement rate details necessary to produce a precise Alcubierre Space-time Manifold under quantized conditions are discussed at length. The reason we are free of the measurement problem with electron-positron pairs is that they are so well characterized.

The measurement problem is averted as the spin states are not compromised by the rapidity of the measurements. Since we are not attempting to establish any significance to Quantum Entanglement or Superposition, the measurement problem does not exist as we are not making any attempt to glean such information out of the measurement, only the spin states.

There are no 'negative energy requirements' necessary to produce the Alcubierre Space-time Manifold, nor wormholes, Casimir effects, 'negative mass budgets,' exotic matter, and so on. This is simple: Alter the progression of time via the laboratory hardened QZE, which in turn alters the shape of local space. There is no need for theories of Quantum Gravity, negative mass, causality violation, reshaping the manifold into a ring, and the many proposals over the years that simply produce a null result, nor any other improvable set of mysteries. These are all unacceptable failures in an otherwise simple approach that requires tabletop, proven technology and no direct energy requirements. To date, no industrial process has produced the QZE beyond the atomic scale. Scaling the QZE up to macroscopic proportions is described.

This work is dedicated in its entirety to Keats Ferrari Bray

All material copyright 2004

"Faster Than Light Travel Using the Quantum Zeno Effect"

Dr. William Joseph Bray

THE TECHNOLOGY

THE ALCUBIERRE SPACE-TIME MANIFOLD

The Alcubierre Metric itself takes the form:

$$ds^2 = -\left(\alpha^2 - \beta_i \beta^i\right) dt^2 + 2\beta_i\, dx^i\, dt + \gamma_{ij}\, dx^i\, dx^j$$

Another form of the metric is written as:

$$ds^2 = \left(v_s(t)^2 f(r_s(t))^2 - 1\right) dt^2 - 2v_s(t) f(r_s(t))\, dx\, dt + dx^2 + dy^2 + dz^2$$

This form is reduced to:

$$-\frac{c^4}{8\pi G} \frac{v_s^2(y^2 + z^2)}{4g^2 r_s^2} \left(\frac{df}{dr_s}\right)^2$$

I understand that this is a common rehash (gratuitous presentation) of the same equation without a detailed examination of the parameters. However, toward the end of this chapter on engine design we will go into the details of this equation at some length and even go as far as to quantize it over the entire 4-dimensional domain and shape of the manifold detail by detail. We will not treat the Manifold as a 2-dimensional rendering of a 4-dimensional façade (Holographic Principle of Quantum Mechanics). Oddly, the Alcubierre Space-time Manifold will not work as a Holographic Quantum Mechanical manifestation. Rather I should say I certainly am not smart enough to make it work as such.

Rather I should say that Holography occurs on a Swarzschild surface where time has infinitely dilated.

Nonetheless, as we shall see somewhat later on, as I address the issue, an equation of Information Entropy expressed as gravitation by Nicolini *forces* the Alcubierre Space-time Manifold from a 4-dimensional façade into a 2-dimensional rendering, AKA, Holographic Model of the Alcubierre Space-time Manifold, quantized down to the Planck scale. This overlooked feature, the fact that the cosmos is not a 4-dimensional façade, but a 2-dimensional Holographic rendering *where time is NOT a valid dimension, but a conscious metered value,* hence, the Holographic model lacks dynamics; has been the major failure of the Physics Orthodoxy since Alcubierre expressed his equation, since Einstein expressed General Relativity.

We will go over it again and again that time, expressed by the same mathematical theorem that is indisputable, can only exist as an infinitesimal, regardless of opinion or seeming 'paradox.' The entire cosmos exists as something that can only be described as a shell, one Planck length thick in any one place. Each Planck unit is completely isolated in 'time' from every other, such that there exists no common frame of reference for any two. What allows for the seeming continuity of space and 'time' will be described a bit later on, and actually has to do with the ill-conceived nature by which we calculate the value, pi.

Later, we will see images from the LIGOs observatory confirming a space-time inversion of the exact same type thought impossible in creating the Alcubierre Manifold *occurring in nature* as two Black Holes coalesce. Where Physicists argued that Alcubierre's Manifold would require most of the universe mass-energy in 'exotic matter' to produce, we see the same inversion occur in nature using a trillion, trillion, trillion, trillion times less mass-energy of ordinary matter.

The common argument is that such a metric requires negative energy, speculatively from an 'exotic matter' source, and the energy requirement regardless of form is on the order of 10^{67} grams of matter or exotic matter, depending on one's speculative definition for negative energy, perhaps a billion times the mass of the universe worth of exotic and/or negative energy. This is common *Orthodox* thinking in today's physics environment. You will see me you the term Orthodox or Orthodoxy often when presenting such awkward approaches to visualizing and/or understanding such concepts in such a non-sequitur fashion, as will be explained shortly.

THE DEMANDS OF SPECIAL AND GENERAL RELATIVITY

It is vital to comprehend these preliminary concepts in order to comprehend the engine's design and working principles. Most of the principles described are novel approaches to Quantum Theory and in some cases corrections to very old errors in the field.

The technology required to construct the working prototype of this engine as described currently exists. Some such technologies will need to be 'modified' or tweaked but no 'new technology' is required. The energy requirements are extraordinarily low, in the kilowatt range, because only detection, measurement, and high-speed computation are required.

As such, *I absolutely guarantee* that the time from proof of principle to demonstrating the prototype can and will be achieved in ten to fifteen years, and not one day more. If one reads through this paper carefully one will find that the actual principle and design of the engine is quite simple and the modifications to existing technology are minimal. Development is almost the total net time cost.

General Relativity dictates that the relationship between massive objects, and the concurrent curvature of space-time (keeping in mind that space-time is a singular construct and the dimensions, although to this day are commonly treated as being separated or isolated in some way, depending on the theory or argument, e.g. 'String Theory, and so on), are by definition indivisible. An alteration in the progression of time **must** result in a curvature of space, and visa-versa, or would otherwise violate General Relativity (not suggesting any causal component). Lately, in an effort to quantize gravity, many efforts have been made to violate and even abandon General Relativity in favor of other phenomenon that will not restrict new, young theorists who cannot work with the rules. However, such Orthodox theories, whereas the classification Orthodoxy is non-sequitur, have no means of testing or providing evidence, they remain hypotheses, e.g., mind games.

Just a warning. Papers written by young theorists today arbitrarily toss out the rules (Special and especially General Relativity, Planck's constant, and so on) in favor of the hypothesis, which isn't even interesting. Somehow, they make it past peer review. Since I'm long retired, I have no idea what darkness favors the foolish. I refer to this as the Dark Side. Any hypothesis that cannot be tested because we are millennia from possessing such technology are referred to as 'theories,' when a 'theory' requires evidence – it thus remains a hypothesis. Thus, there is no such thing a 'String Theory,' 'm-Theory,' and so on, these possess no evidence that we will find a means of detecting for centuries to come, longer if they do not exist. They are hypotheses. String Hypothesis, m-Hypothesis, and so on.

According to General Relativity, in simple terms, we measure time as progressing slightly slower on Earth's surface than in space. There is no provision for causality in this relationship in General Relativity. That is, it is non-sequitur that a curvature of space *results in* the slowing of progression of time and that the slowing of the progression of time will *not result in* a curvature of space; i.e., there is no causality, neither is the 'result' of the other. Furthermore, there is no evidence to support that a slowing of the progression of time *will not* 'result in' a curvature of space. The relationship has no causal component, nor is there any provision or precedence for a lack of causal component. That is, until now, we have had no way of artificially manipulating either space or time. Now we can produce a suitable QZE. The reason we have not seen an accompanying curvature of space to date is that it has only been carried out on a subatomic scale.

The goal then is to scale the phenomenon up to produce a causal relationship in General Relativity resulting in a concurrent curvature of space. If a concurrent curvature of space does not occur, then General

Relativity will be violated, and this will be a greater enigma then producing the first artificial curvature of space.

However, there is no provision that allows for one to occur in nature or otherwise without the other, no provision that allows for a slowing of time without an accompanying curvature of space. As time slows space curves, as time speeds to space-normal space flattens back out again. Those are the simple rules of General Relativity, there is no evidence we have seen anywhere in any telescope of any sight in the cosmos that these rules are bent or broken. We use these rules to measure space and things we cannot even see on the other end of the universe. We have actually weighed the mass-energy of the cosmos with this simple rule, seen otherwise invisible Dark Matter and Dark Energy that make up 95% of the mass-energy of this cosmos.

An alteration of the progression of time *must* be accompanied by a curvature of space, or will otherwise violate General Relativity.

In simple terms, as many regard the QZE as a means of altering the flow or passage of time, it is simple to regard it as the alteration in the shape of space. For this, we turn back to our arrow in flight. Eyes on the arrow, it is frozen in flight; the passage from one Planck interval of space, 10^{-35} meters to the next becomes infinite. We have altered the shape of space, actually made the path of 'unitary space' infinitely long. We blink our eyes, the arrow makes it to the next frame, and we have again altered the shape of 'unitary space,' providing continuity from one frame to the next.

We can take this to less extreme and simply alter the progression of time by a factor of two, which in turn alters the path of the arrow by a *relativistic* factor of two, and so on (it is not a 1:1 relationship). The QZE in this manner has obeyed the laws of General Relativity all along unnoticed, because of the scale it has been operating at.

In Special Relativity, again, there is no causal relationship between an alteration in the progression of time and the accompanying alteration in the shape of space, only an ***implied*** causal relationship that both the alteration of the progression of time and the alteration of the shape of space co-consequently result from velocity, provided this observation is from some chosen preferential perspective (frame of reference, Special Relativity). Again, it can be stated that an alteration (in Special Relativity) of the progression of time ***must*** be accompanied by an alteration in the shape of space as observed from this chosen frame of reference. In General Relativity, this preferential perspective is placement with respect to a place within a gravity well (in theory a gravity well extends for infinite distance, thus everything is within everything else's gravity well). In this case, distance from the *mean center* of the phenomenon becomes our preferential perspective.

THE MATHEMATICAL CORRECTIONS AND NOVEL APPROACHES TO QUANTUM MOTION: EQUATING THE QZE WITH TIME DILATION

The Quantum Zeno Effect is defined as an alteration in the progression of 'unitary time' by rapid or constant observation. In this sense, 'unitary time' was originally intended to refer to the progression of one Planck interval of time to the next (one 10^{-44} second 'snapshot' to the next). 'Time dilation' is typically thought of as a separate phenomenon where the progression of time is altered as the result of some referential and/or preferential perspective either as a result of velocity (Special Relativity) or gravitation (General Relativity).

As I have pointed out, the QZE *must* be an alteration in the progression or shape of 'unitary space.' The QZE is in fact an alteration in the continuity of unitary space and unitary time.

In order to equate the two phenomena, the progression of unitary time (the progression of one Planck interval of time to the next) to time dilation either as a result of the conditions of Special or General Relativity, one takes into account the preferential perspective of the observer, which to date seems to have been ignored but is otherwise obvious.

In Special Relativity the observer is, for example, stationary and observing an object moving at high velocity toward or away from the observer. Regardless of what is actually occurring regarding the observed object the observer, being stationary, can only measure changes of the observed object with a 'meter stick' which is quantized; because it is both stationary and therefore quantized. It is not possible to measure a value of the observed object resulting in a non-quantized value. (The phenomenon of 'quantized red-shift' observations of distant galaxies is a perfect example, but it hasn't occurred to anyone what the apparent cause of this phenomenon is aside from what is stated here, but is otherwise obvious once stated).

That is, with respect to Relativistic changes, a stationary observer has a meter stick that is quantized to his/her stationary frame of reference, and it is therefore not possible to measure any non-quantized value of an object moving at Relativistic velocity. Special Relativity is therefore *Quantized* according to Planck time, t_p, Planck mass, m_p, and Planck length L_p in all frames of reference.

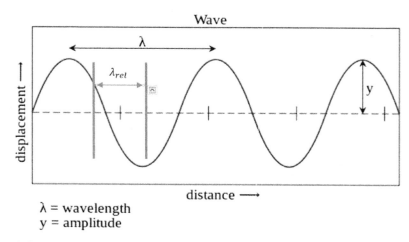

λ = wavelength
y = amplitude

In this depiction, lambda represents one Planck length, and as such would characterize our locally quantized meter stick in our stationary frame of reference (Special Relativity) or local gravitational frame of reference (General Relativity). The lambda$_{rel}$ is an impossible value because it is not an integer value of a Planck interval. It represents what happens to the Planck length of an object at relativistic velocity or gravitation when we attempt to measure it with our locally quantized meter stick. The two values, each the standard Planck length, do not concur. Something has fundamentally changed, and it is not merely an observational sleight of hand. A real change in the fundamental length, Lp has occurred. However, it has occurred in all frames of reference, since the observed (under Special or General Relativistic conditions) meter stick is also quantized to their local environment.

Everyone's meter sticks are quantized to their local environments. No one can measure another's condition as any other value than a quantized integer value of Lp or tp. Since it is agreed that these are not merely 'observed' effects, but real, the logical, sequitur conclusion is that Lp and tp change. Furthermore, we do not require extreme velocity or extreme gravimetrics to do so, since everything is in motion, and everything has mass, it is therefore axiom that Lp and tp are quantized to as many local conditions as can be defined, on every scale.

Again, we address the issue that some theorists argue, albeit without evidence or compelling reason, that the Planck Length, Lp is not the fundamental length allowable in this cosmos. So we look at the equation for Lp, whose derivation is not debated:

$$L_p = \sqrt{\frac{hG}{2\pi c^3}}$$

In the denominator we have 2 pi c, all rock solid constants, albeit pi is a bit fuzzy. In the numerator we have G, which is considered solid, and h. Lp is entirely dependent upon h, the fundamental quanta of this cosmos. If, then ,else: If Lp is not the fundamental length of this cosmos, then h is not the fundamental quanta of energy in this cosmos; else, Lp is the fundamental length in this cosmos and h is the fundamental quanta of energy in this cosmos. Let's keep that perfectly clear.

Now for the fundamental argument, in which way does Lp change, larger or smaller? History dictates that length contracts in the direction of travel at high velocity, but as we shall investigate at depth, that equation has been upside down for over a century, and we will also investigate why, and why it is absolutely impossible for that equation to be correct as historically written.

In an *accelerating body*, moving away from the observer, for instance, it is not possible to observe a 'smooth' transition in the time dilation effect of the observed object because all such measurements taken from the stationary perspective are quantized according to the stationary observer's quantized 'meter stick.' It can be stated as an axiom that the quantization of the stationary observer's meter stick does not become 'un-quantized' as a result of the observation of an accelerating body. Historically, the idea that time dilation in Special or General Relativity were quantized has never occurred to anyone, it has always assumed to be 'smooth.'

The same principle holds true for gravitation. The measurement of the curvature of space can only be quantized because regardless of the condition of the massive body and its curvature, there can be no 'smooth' measurement of such curvature because the observer can only use a quantized meter stick within their gravitational reference to make such observations. Again, it can be stated as axiom that the observer's quantized meter stick does not become 'un-quantized' as a result of measuring the curvature of space around a massive body.

Because the observer taking the measurement is using a meter stick that is quantized according to the conditions of their local environment, and it is not possible that their meter stick is not quantized to their local environment, there can be no measurement or observation of any system under Special or General Relativistic conditions that is not quantized. It does not matter if the measurement in question in Special Relativity is by the stationary observer, whose meter stick is quantized to their local environment, or the traveler, whose meter stick is quantized to their local environment. In General Relativity, again, it does not matter if the measurement is taken by an observer at great distance, whose meter stick is quantized to their local environment, or by an observer deep in an intense gravity well, whose meter stick is quantized to their local environment.

Time dilation, length contraction, and so on, as a result of velocity (Special Relativity) or gravitation (General Relativity) can therefore be stated ***as an axiom*** as being ***quantized***. In addition, there is no way around this argument. There is no possibility of taking the measurement, by any observer in any frame of reference, other than via a quantized 'meter stick' within any observer's local (Special Relativity) or gravitational (General Relativity) frame of reference, whatever that may be, and regardless of scale, from the Planck scale out to cosmological distances.

The argument that velocity is quantized does not work. For instance, 0.5c is one Planck unit of length divided by two Planck units of time:

$$1L_p/2t_p$$

The other possibility being **½ L_p/1t_p**, but this is not a possibility in normal space-time:

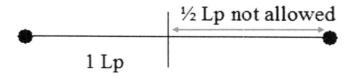

At 0.5c, **½ L_p/1t_p** is not an option because ½ L_p is not a possibility in normal space-time.

Moreover, this value, $1L_p/2t_p$ is unambiguously quantized. However, the resulting time dilation result goes out to at least several hundred decimal places (thousands on my arbitrary precision computer algorithm), exceeding the orders of magnitude of our Planck values, and truncation then becomes arbitrary. Therefore, it is not the quantization of velocity that is responsible for the observed quantized time dilation. The observed time dilation can only be quantized, and only as a result of the observer's stationary, quantized, meter stick as measured from the chosen stationary frame of reference.

That is, velocity is nL_p/xt_p in terms of Planck scale, and can only occur in integer values of n and x.

Furthermore, it is *axiom* that the observed phenomenon, be it a measurement taken within a speeding body (Special Relativity) or measurements taken within a massive body (General Relativity) remain quantized from the preferential perspective of being at the center of the phenomenon. That is, a person measuring their own state at high velocity, accelerating, or in a gravity well, will still be using a 'meter stick' that is quantized within their local system. The same holds true for every observer according to the conditions of their local system, everywhere.

There is no frame of reference under any set of conditions where a 'meter stick' used for the measurements of space-time are not quantized. By 'meter stick', I mean that to include metering the progression of time or mass as well, or any other measure.

It is therefore *axiom* that the alteration of space and time as a result of velocity (Special Relativity) and/or gravitation (General Relativity) are equated in this first case to the Quantum Zeno Effect in that they are all quantized and the result of observation. There are, in fact, no other conditions that exist necessary to qualify any difference between time dilation as a result of Special or General Relativity or the QZE. Whereas we are accustomed to seeing Special and General Relativity on macroscopic scales, the QZE thus far has been limited to subatomic scales.

The exact quantization of the QZE will be described a bit later in this paper.

Whereas William Crooke established the first stable electron beam in 1869, it can be said electrons 'rule the world' today. Once a technology is scaled up, replication is merely a matter of production.

Neither higher mammal nor nature can differentiate them, Relativity and the QZE, and they are therefore equivalent, regarding that there cannot actually be any 'gaps' between Planck intervals of time to explain or allow for non-quantized variations, rendering the obvious conclusion that an effect has occurred altering the *real* Planck interval of time (and space) in all frames of reference.

That is, the changes do not occur because there is any 'play' between Planck intervals as though they were marbles spaced out on a surface to be squeezed or separated.

The observed frame of reference has undergone a *real change in the Planck interval* according to the observer, and the observer has undergone a *real change in the Planck interval* according to the observed, in both Special and General Relativity. Just as mass must undergo a *real change* in a particle accelerator or such a thing would be useless technology, mass, space, and time are bound by countless equations and undergo real changes under Relativistic and QZE conditions.

That is, clearly, since my meter stick is quantized according to my local conditions, and the 'observed' either at high velocity (Special Relativity) or deep in an intense gravity well (General Relativity) has their meter stick quantized to their local conditions; it is clear that the value for the Planck interval, L_p has changed according to any observer's frame of reference. This would then of course be true for the Planck interval of time, t_p.

This also explains the failure of the Michelson-Morley experiment, detailed further on.

Furthermore, since space and time are indifferentiable, space-time, the *real* Planck interval of length has also undergone a change in all frames of reference.

We simplify this exhaustively difficult argument by turning back to our classic 'Twin Paradox.' The speeding traveler observes the stationary observer as 'speeding up' (taking less rapid measurements; the photons as viewed from head-on are shorter, allowing less time to observe each photon according to the observer's watch). The stationary observer sees the speeding traveler as 'slowing down' (taking more rapid measurements of the speeding traveler; the photons as viewed head-on are longer, allowing more observations of each photon according to the observer's watch). On a quantum scale, however, each bit of information observed by the stationary observer, in this case a photon as measured from its beginning point to its end, has taken a longer path, at least at the moment of observation.

In this entire argument we negate the typical incorrect application of acceleration being the cause of differentiation between the observer and observed, and use the correct application of only Lorentz *variation of length* and Relativistic Doppler Effects.

The fact that this 'longer path' from beginning to end of the photon in question is a wave function crossing the observer's stationary or preferential frame of reference indicates that the observer is *interdependent* with the observed photon. That is, *'the longer path' in question is in the observer's frame of reference.*

This is the QZE and is equated to both Special Relativity and General Relativity and the argument holds true in a gravity well.

We then extend this argument to include the *observed effect* of length contraction by Michelson-Morley's interferometer *in all directions* (on a Planck scale):

$$Lp' = Lp\sqrt{1 - (\frac{v}{c})^2}$$

The Michelson-Morley interferometer 'got seemingly shorter' as a result of the *real cause:*

$$Lp' = \frac{Lp_0}{\sqrt{1 - (v/c)^2}}$$

Space-time is expanding in all directions. *However, this wouldn't be discovered for another half-century.* In any general sense, however, no detector can measure any relativistic transformation of its own condition, regardless of whether the equation is upside down or not; *and that is why the Michelson-Morley experiment failed.* Yet, there have been perhaps fifty or so attempts since Michelson-Morley, each using ever increasing technology, the latest in 2009! Moreover, they are still at it, all null results, with accuracies out to 10^{-17}, with complete disregard for the most basic tenet of relativistic transformations; *you cannot detect your own relativistic transformation in your own frame of reference.*

Again, length contraction has never been experimentally validated, and the experiment that fueled the entire argument, the Michelson-Morley experiment, proves exactly the opposite.

As will be shown later, there is no direct or implied experimental evidence to support length contraction as a result of Special or General Relativity. Although this may come as a surprise to some, given that time dilation is extremely well characterized, to date, no one as even proposed an experimental approach that is capable of determining length contraction. As will be shown below, the only such experiments were performed half a century ago and yielded a result of length dilation, not contraction, both in Special and General Relativistic conditions. There is some anecdotal references to the possibility that some particles (in particle accelerator experiments) appear to behave as though they were more 'plate-like' at high velocity, but admittedly speculative and non-conclusive.

The most damning piece of evidence is our upward spires emanating from the collision of two Black Holes clearly indicating a space-time inversion (lengthening) of a sort that is on the order of three or more solar masses:

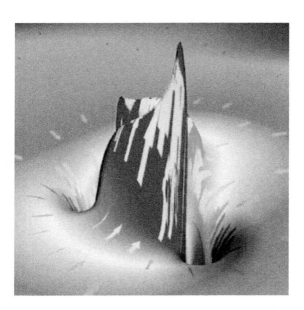

Here we have the energy of the collision causing a space-time inversion (lengthening) that will flatten out to produce gravitational waves carrying off energy of roughly three solar masses that will be detectable for billions of light years distance.

There is also an issue of 'semantics' here. In one sense, we say that according to the photon the distance travelled is zero, and call that length contraction (in some circles of argument). However, since the travel time to any point on the cosmos, to a photon, is zero, then we can argue that the photon simultaneously occupies every point in space-time in the same Planck instant (Super-positionality). From the photon's perspective, its own length, size, what have you, is *infinitely dilated to extend to every corner of the cosmos, and only a special event, once termed 'Wave Function Collapse' will determine the final destination of the photon. In this sense, the photon is considered dilated to infinite length.*

LORENTZ EQUATION

It is also of importance to note that using the hard definition of the speed of light, particularly on a quantum scale (but is also true on a macroscopic scale):

$$c = 1L_p / 1t_p$$

Then

$$c = L'/t'$$

The conditions that time dilates:

$$t' = \frac{t_0}{\sqrt{1 - (\frac{v}{c})^2}}$$

But length contracts:

$$L' = L_0 \sqrt{1 - (\frac{v}{c})^2}$$

Requiring that:

$$c = L'/t'$$

Results in:

$$c = \frac{L_0 \sqrt{1 - (\frac{v}{c})^2}}{\frac{t_0}{\sqrt{1 - (\frac{v}{c})^2}}}$$

$$c = \frac{L_0}{t_0} \left(1 - \left(\frac{v}{c}\right)^2\right)$$

When $v = c$

$$c = \frac{L_0}{t_0} (1 - 1^2) = 0$$

This requires that the speed of light is variable *in all frames of reference, if Lorentz contraction is true and correct.* That is one bleeping upside-down equation. When shown in this fashion there is no mathematical redemption for the concept of length contraction.

In fact, this form, which uses the classic arrangement of 'Lorentzian contraction,' states that an object speeding away at the speed of light is not moving at all in any frame of reference (when v=c, v/c=1, the term $(1-(v/c)^2)$ becomes zero, and the speed of light denoted by 'c' would then be zero. This is as simple as the photons leaving your flashlight being at a dead-stop. The math is straight forward and irrefutable.

However, if we regard length as dilating:

$$l' = \frac{l_0}{\sqrt{1-(\frac{v}{c})^2}}$$

In addition, time as dilating:

$$t' = \frac{t_0}{\sqrt{1-(\frac{v}{c})^2}}$$

The, regarding the hard definition of the speed of light on a quantum scale (also true on a macroscopic scale) that:

$$c = 1L_p/1t_p$$

Then:

$$c = \frac{\frac{l_0}{\sqrt{1-(\frac{v}{c})^2}}}{\frac{t_0}{\sqrt{1-(\frac{v}{c})^2}}}$$

And

$$c = L_p/t_p$$

Again, the math is straight forward and irrefutable. The very simple math is quite irrefutable in dictating that Lorentz equation as classically depicted has been upside down for over a century. Lorentzian variation is in the form of dilation, not contraction.

If one sits down and analyzes the Twin 'Paradox' with Lorentzian Dilation rather than contraction by merely flipping the fractions the Twin 'Paradox' is resolved in exactly the same fashion with disregard to acceleration and other incorrect approaches. That is, an object travelling at high velocity extends, or as I say, 'stretches' its nose toward it target, superpositioning itself as Quantum Mechanics demands. The seeming 'paradox' is having the entire shape of the cosmos change, again, for a single photon's path. Again, the word 'paradox' literally means; *absurdity*.

That is, with some thought, the absurdity arises in the mere concept that the passage of a single photon contracts the entire QED vacuum if a billion light-years of space-time and the entire mass of two stars it passes between… If we regard the QED calculated value of 10^{113} joules per cubic meter that's 10^{137} joules just for the empty space (10^{125} [more than a googol] Hiroshima bombs) plus the mass-energy of the stars…

Or is it less absurd to consider the single photon as superpositioned between two points in space-time by extending Lp via length dilation? For arguments sake, take the Twin Paradox, and just turn every length contraction into a dilation, and see that you get the same result. The absurdity is in finding that the argument has been upside down for a century.

For General Relativity, requiring that the speed of light is non-variable in all frames of reference, as is the known case, and with similar results when applied to length in a gravity well if we use length *contraction* as an argument in a gravity well, according to:

$$L' = L_0 \sqrt{1 - \frac{2GM}{rc^2}}$$

However, time dilates:

$$t' = t_0 / \sqrt{1 - \frac{2GM}{rc^2}}$$

The result is:

$$c = \frac{L_0}{t_0}(1 - \frac{2GM}{rc^2})$$

As explained earlier, I use t' in this form to be consistent with t' in Special relativity. In either case, it doesn't matter, if one dilates and the other contracts, the equation still ends up in this final form.

This leaves the speed of light variable. In simple terms, as M or r (the only two variables) changes in this equation, c, the result changes. This is not a misplacement of time dilation equations. The dilation factor has been eliminated and as you can see, L_0 and t_0 are being held constant, leaving only c as a variable. Also, this is not a case of red-shifting out of a gravity well, this is a change in c as a velocity, as it is on the left hand side of the equation.

Although this seems intuitively to be the case, it cannot be the case. The primary variability in question here would be M, the center mass, such as a planet or star, and r, the distance from M. In this case, instead of light bending along the curvature of the gravity well, it speeds up and slows down as though the photon had mass. The variables L_0 and t_0 are spectator variables.

By flipping the equation upside down, in its proper form (length dilation in a gravity well):

$$l' = l_0 / \sqrt{1 - \frac{2GM}{rc^2}}$$

In addition, time dilation in a gravity well:

$$t' = t_0 / \sqrt{1 - \frac{2GM}{rc^2}}$$

Again, using the hard definition that c=1Lp/1tp the result is:

c=1Lp/1tp

The answer is length dilates in a gravity well with respect to a distant observer. This is consistent with the fact that the approach to a Schwarzschild radius is asymptotic, requiring infinite time and distance, and why I say that a Black Hole never forms but is always in the making.

In the simplest sense, we can think of the passage toward the event horizon, or rather, the Swarzschild Radius, of a Black Hole to be asymptotic. Even the collapse of a massive body towards the Swarzschild Radius never reaches the Swarzschild Radius, thus a Black Hole is forever in the forming, but never reached. Nevertheless, the passage of anything toward this endpoint can said to be of *infinite distance*, because that approach is asymptotic.

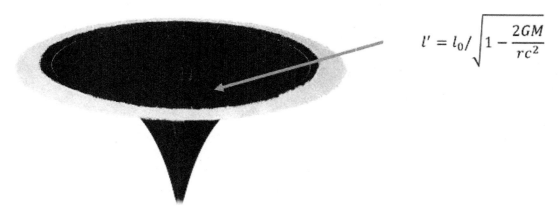

$$l' = l_0 / \sqrt{1 - \frac{2GM}{rc^2}}$$

Since space and time are inseparable and interdependent elements of nature, it is not possible to have time dilating and space contracting. They must both do the same thing:

$$c = \frac{L_p}{t_p}$$

Thus, length contraction becomes an absurdity of the Orthodoxy, a belief without data, disproven by elementary algebra. Furthermore, as stated, General Relativity forbids that space and time do opposite things. That is, time dilate while space contracts, regardless of the confusion of relativistic transformations occurring, they, space and time (space-time) ***must*** do the same thing, even during a relativistic transformation, otherwise violate General Relativity.

$$c = \frac{Lp' = Lp_0 / \sqrt{1 - \frac{2GM}{rc^2}}}{tp' = tp_0 / \sqrt{1 - \frac{2GM}{rc^2}}}$$

Which simply reduces to:

$$c = \frac{l_{p\prime}}{t_{p\prime}}$$

The understanding of these principles is vital to comprehending the structure of the Alcubierre Space-time Manifold. As we shall see, there are no exotic conditions, and there is no actual motion taking place. The Alcubierre Space-time Manifold works by superpositioning the vessel at the start and endpoints via length dilation of cosmological proportions, regardless of the founder's original intent, a product of Orthodox thinking.

That is, the Alcubierre Space-time Manifold superpositions the vessel from the starting and destination simultaneously. It is a superpositioning manifold, not a Faster Than Light AKA warp drive in the sense that Alcubierre or physicists have interpreted it as since it was first published in 1994. The drawback to comprehension was the upside-down thinking of 'contracting space.'

Keep in mind that to date, in simple terms, the entire premise for the form of the Orthodox General Relativistic Lorentz equation being:

$$Lp' = Lp\sqrt{1 - \frac{2GM}{rc^2}}$$

Leaving a maximal result of Lp'= 0. Since a Schwarzschild radius can never be reached because it is an asymptote, the value zero is an impossibility, far below the Schwarzschild radius.

The proper form should be

$$l' = l_0 / \sqrt{1 - \frac{2GM}{rc^2}}$$

With a potentiality of L'=∞

Is based upon the simple math:

$$v = \frac{L}{t}$$

In this equation, 'v' is a variable. In relativistic conditions, c is an absolute invariant. The value c is not only an absolute invariant; it represents a domain where time does not exist. Even if you want to stick with that classical argument, at velocity c, $\Delta t = 0$, *always,* in which case:

$$\Delta t = 0; \frac{L}{0} = \infty$$

Which describes the state of the photon at all times. From the photon's frame of reference, all travel times to any distant locations are zero, and its own length is Superpositioned throughout space-time to infinite distance (this is in line with Schrodinger's equations and Von Neumann's Copenhagen Interpretation of Quantum Mechanics).

The velocity of light, both in Special Relativity and General Relativity are constant in all frames of reference, *if and only if we regard length as dilating, not contracting,* as we have seen in the simple algebraic manipulations above. In both cases where the mainstream equation for length contraction is applied, the speed of light *varies in all frames of reference.* Moreover, noted the mainstream rationale is *an observed property, not a real change in Lp or tp.* Historically this is because Newton defined the gravitational constant, 'G,' as *a constant.* He of course had no means available to prove this; it simply worked in observing planetary orbits. No one has ever challenged it, and by some 'hail Mary' approach 'G' is regarded as fixed and immutable, although that has never been tested under relativistic conditions.

The 'real' change in both Lp and tp are the cause of the observed effect. As explained earlier, as Lp increases, the unit time to cross a greater distance is observed as a decrease in length; as a result of a 'real' increase in the Planck length, Lp.

The notion that Lp or tp can be regarded as variable will sound unacceptable to some. However, there is no evidence to date that indicates, much less proves, that they are fixed and invariant. They have been historically treated as such, as a matter of convenience, otherwise accounting for the variations would add renormalizations and complex characterizations that need to explain the invariant velocity of light while simultaneously explaining the variance in the Planck values of time and length. However, as shown above, I have just provided that. It is merely a correction to a key variable (length contraction vs. length dilation) that was historically upside down (to resolve the Michelson-Morley experiment) as a result of expanding space-time, unknown to Lorentz who considered length contraction, half a century before expanding space-time would be discovered. As stated, our variation in Lp and t_p is founded upon our locally quantized meter sticks.

That is, to date no one has explained satisfactorily what 'expanding space-time' is. In fact, no one has satisfactorily explained what space-time is.

The approach to 'fixed' values of Lp and tp are based on the assumed fixed values of the constants they are composed of:

$$Lp' = \sqrt{\frac{hG'}{2\pi c^3}} \qquad t'_p = \sqrt{\frac{hG'}{2\pi c^5}}$$

The rationale for G' will be explored shortly.

For the record, Lorentz nor his colleagues never resolved the issue, 'why the Michelson-Morley interferometer contracted in every direction it faced.' When Einstein provided his proof for Special Relativity he clearly showed that only the 'x' direction applied in a 3-d coordinate system of {x,y,z}.

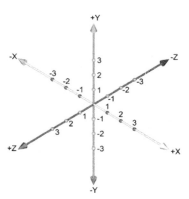

Lorentz ignored this, or otherwise overlooked it. In Special Relativity, length is only affected in the direction of travel, 'x.' The height and width remain unaffected. Thus, Lorentz must have used the same mathematical approach that Einstein did, meaning that he knew only the 'x' direction would be affected. Michelson and Morley turned their interferometer in every direction, they turned the table on its side and then turned the interferometer in every direction, knowing that the planet was spinning the whole while. Still, they got nothing. Because Lorentz was a mathematical giant, whatever he proposed was therefore accepted.

Lorentz presented a mathematical solution; the matter was otherwise unresolvable and forgotten. However, if one looks at the Lorentz explanation, the question begged for an answer. Why, according to Lorentz 'contraction' equations would the interferometer 'contract' in *every direction* it faced? Keep in mind this is not on a flat plane, and it wasn't clear to all that Lorentz transformation was only suited for the 'x' vector at the time it was presented as a solution for the Michelson-Morley experiment. The experimental setup is on a bench top on a planet that is rotating at 1000 miles per hour, tilted on some non-referenced axis, orbiting the sun, which is orbiting the galaxy, and so on. It was in fact designed to measure the velocity through the 'ether' around the sun and failed even facing that direction. Lorentz solution provided an answer that was off by a factor of over 100. Yet, historically it was accepted and the matter dropped. To this day it is still taught as the Lorentz solution.

In simple terms, even if the interferometer were facing some fixed point it is spinning and turning head over heels wildly in every plane and axis at a high rate. The simple answer is that Lorentz could not know that half a century later it would be determined that space-time is expanding in all directions. In his day, the universe was the Milky Way, fixed and static. This however is not an explanation for the null result. The null result is the tenet that you cannot detect a change in your own transformation under relativistic conditions.

That is, clearly, your meter stick is quantized to your local environment. Although we discuss this in high-school, super accurate versions of the Michelson-Morley experiment are still being tested to this day to see if that statement is true.

My entire rationale for approaching this topic at all is that as one 'scientist' pointed out his inability to figure out which way Alcubierre's Space-time Manifold was meant to travel by looking at its *symmetry*. At the same time I do not want to design an engine that goes several hundred times the speed of light *backwards* based on obsolete 19th century equations and handed down thinking. That is, thus far we have had to dismiss a trillion, trillion, trillion, trillion, trillion universes worth of 'negative energy,' in order to *contract the QED vacuum* from here to the nearest star (travel 4 light years of contracted carpet to move an inch), in a variable 'c' environment where the speed of light equals zero, using an Alcubierre Space-time Manifold that is upside down and backward, that requires a chemical rocket 'boost' to cross the galaxy at thousands of times the speed of light (which is zero) in order to satisfy the physics Orthodoxy. And this is not a joke, the claimant in this case is currently funded by the U.S. government to research this 'stuff' with the claim; 'it may take two years, it may take two hundred years...' I call that 'fired.'

The simple algebraic manipulation of these values to their proper form resolve innumerous paradoxes and also agrees with the only hard data regarding the measurement of length under these conditions; keeping in mind that 'length' refers to things, particularly things with mass, and not only space-time itself. However, I am not going to, at this time, negate space-time as having substance. I might, in fact, substantiate such an idea at some point. When I say 'space-time' has no length, nor does it have time, I am referring directly to the Holographic Principle of Quantum Mechanics, which reduces our 4-dimensional façade to its true form of a 2-dimensional Swarzschild surface. We will discuss that briefly later.

The argument that any of these effects are only 'observed' effects is non-sequitur. All of these equations are derived from the same set of equations under the same principles as:

$$m' = \frac{m_0}{\sqrt{(1-(\frac{v}{c})^2}}$$

The entire working principle of a particle accelerator is based upon this equation. If the mass increase were not 'real,' the particle accelerator would do nothing. The 'real' mass increase is required in order for the accelerator to produce heavy particles (along with a spray of lighter particles whose sum is greater than the starting mass).

The Alcubierre space-time manifold, then, is classically drawn both upside down and backward, *if one references the historical logic of space-time contraction*. My depiction of Alcubierre's manifold is correct, although most would regard it as depicted upside down and backward. I am therefore sticking with the argument that I perceive as obvious, and depicting the manifold and describing it as being *corrected*. *However, if one ignores the historical upside-down references to space-time contraction the depiction of the manifold is arbitrary, one needs simply to agree on what the graphic depiction represents.* There are many more corrections to the manifold I will present later.

The Alcubierre space-time manifold as classically drawn is the result of having Lorentz equations upside down. The result is an erroneous supposition of the need for 'exotic energy' to manifest the space-time manifold. As shown from the LIGOS data, exotic matter and/or negative mass-energy are not needed to produce a space-time inversion:

In this LIGO model of coalescing of two Black Holes of approximately 30 solar masses each (Confirmed by LIGO data of detected gravitational waves), the two downward wells are typical space-time wells at the base at either side of the spires. The two upward spires occur for a few hundred microseconds, elevating above the flat background of space-time, they are *a space-time inversion*, occurring in nature. Be it understood that the conditions under which they occur are *intense*; they are not exotic, such as negative energy and so on. Such an inversion was thought to be an impossibility, limiting the potential to manifest an Alcubierre Space-time Manifold. This space-time inversion, occurring as shown here in nature, is exactly what is described in Alcubierre's manifold of you flip it such that it is not a Lorentz blunder.

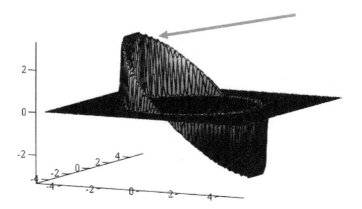

When the Lorentz equation is corrected, the need for 'exotic energy' is dismissed. The curvature of space-time to produce the manifold then becomes the direct result of detection and measurement at very high speed the Quantum Zeno and Quantum Anti-Zeno Effects), and nothing more. Granted, the LIGOS data is of two coalescing Black Holes, but as I have stated, the Quantum Zeno Effect is capable dilating time to zero, whereas a Black Hole cannot. Assuming it can be scaled up, and I will blueprint the mechanism for doing so in this paper, 'painting,' as I refer to it, an Alcubierre Space-time Manifold becomes obvious, once explained.

Also note that Einstein's original (hand written) paper later regarded as 'Special Relativity' inexplicably flips Lorentz equation as classically depicted as upside down {his original handwritten notes are available from the Jewish-American archives in Washington D.C., not online} with no explanation in order to derive the correct equation for time dilation. It is likely Einstein perceived the error and merely used the correct form, stating nothing, as a *patent clerk* does not point out errors in a mathematical *rock star*. Furthermore, his later *hand written* works depict Lorentz equations as flipped; indicating he consistently regarded the error but said nothing, probably for trepidation of quaking the Orthodoxy. Einstein consistently remained in line with the Orthodoxy, yet had no good thing to say about them. There are several quotes where Einstein seems rather bitter regarding the Orthodoxy in math and physics. In all likelihood he spent his stellar career walking on eggs regardless of his status.

In any case, it is not possible for the equation to be correct as classically depicted while maintaining light as invariable in all frames of reference, as we discussed a few pages back. This is indicated by the very simple algebra presented above. The 'opinion' that this is an observed effect violates the law that the speed of light is invariant *in all frames of reference. However, the observation; again, the Michelson-Morley interferometer was observed to decrease in length (Lorentz 1886) by what should have been 500 nm, but Lorentz equations only provide 5 nm, and Special Relativity forbids any detection of the transformation of one's own state altogether. In addition, that principle has eluded mainstream physics for a century. It is, in fact the only valid experiment in Lorentzian physics and validates space dilation, not contraction.* That is, since the interferometer in question was well suited to detect the 5 nm predicted by Lorentz equation but failed to do so, the interferometer must have expanded by 5 nm in order to make up the missing length. However, that would violate the tenet of detecting one's own state under relativistic conditions.

Furthermore, again, the effect is real, not observed, just as the mass increase in a particle accelerator must be real, not an observed effect, or the entire system will not work. This in turn requires that time dilation is a real, not an observed effect.

I quote:

Length contraction was postulated by George FitzGerald (1889) and Hendrik Antoon Lorentz (1892) to explain the negative outcome of the Michelson–Morley experiment and to rescue the hypothesis of the stationary aether (Lorentz–FitzGerald contraction hypothesis)

1. FitzGerald, George Francis (1889), "The Ether and the Earth's Atmosphere", Science 13 (328): 390, Bibcode:1889Sci....13..390F, doi:10.1126/science.ns-13.328.390, PMID 17819387

2. Lorentz, Hendrik Antoon (1892), "The Relative Motion of the Earth and the Aether", Zittingsverlag Akad. V. Wet. 1: 74–79

Here, it is noteworthy that the Michelson-Morley experiment was originally set up in anticipation of testing their hypothesis on detection of the motion of the Earth in relation to its path in orbit around the sun, which it failed to do when rotating it 360 degrees on all planes. Lorentz contraction cannot explain this. The experiment was tested to detect motion in relation to other planets, and so on, rotated in 360 degrees in all planes, and failed to do so. Again, Lorentz transformations cannot explain this. This is not because of a lack of 'aether.' The velocity corresponds to theoretical velocity around the sun of about 30 Km/s. This velocity makes no sense in either case, contraction or dilation. That is, in the case of contraction, why the interferometer failed at a full 360 degrees of rotation is not length contraction in the 'x' vector.

The Michelson-Morley experiment, based on a velocity of 30 Km/s around the sun expected a shift of 500 nm based on their 11 meter interferometer. The Lorentz contraction theory was readily accepted to account for this, and has remained as such for well over a century unquestioned. However, at 30 Km/s, the shift is only 5 nm according to Lorentz equations, which is still within the detectable range of their interferometer, yet also remained undetected. Again, Orthodox physics readily agreed on the premise, which is taught as Orthodox Canon to this day, and no one has done the math.

The very premise of relativistic changes is that the traveler, in this case the interferometer, detects no changes in his/her/its own state in their own frame of reference of any relativistic transformation. Yet, again, it is taught as Canon in physics that the interferometer detected a change in its own relativistic transformation in its own frame of reference at 30 Km/s around the sun. And again, it has been repeated so many times for so many years that no one questions it, even though the experiment has been repeated until a recent sensitivity of 10^{-17} meters with a null result. Why the most basic tenet of not being able to detect one's own relativistic shift has been abandoned in these tests is questionable. They are not tests to prove that one cannot detect one's own relativistic shift. They are tests to detect a relativistic Lorentzian shift in length. I am truly amiss.

I am not ranting. The reason for going on about these issues is that as I have stated, the reason we are *not already travelling beyond light speed now* is because of Orthodox thinking. Thinking outside of this box is not at all difficult. *In my opinion,* the box is rather absurd.

As for the QZE, after the resulting curvature resulting from measurement rate is characterized by experimentation and development, the rapidity of the measurement is then 'mapped' back into the emission and detection rate necessary to 'paint' the manifold as depicted (the equations for doing so are detailed later in this paper). Keep in mind, again, that my depiction of the manifold is both upside down and backward of the classic depiction of the Alcubierre space-time manifold, because the classic depiction is based upon the inverted Lorentzian equation as an observed effect, not a 'real change' in the shape of space-time. If a 'real' space-time manifold is to be produced, the Lorentzian equation has to be flipped.

The correct approach is to extend the nose of the craft to the target ($L' \rightarrow \infty$), length dilation, not contraction, in this case our distant star, and snapping the rear of the craft back to where the nose is at our designation

($L'=L_0$). As stated, length contraction is non-sequitur in a cosmos that by definition exists only as an infinitesimal. Length dilation, however, is sequitur to the argument.

When the Lorentz equation is corrected and the Alcubierre manifold flipped right side up and forward, this is exactly what the manifold will do, extend the nose of the craft to the target (increase Lp to some very large value), and snap the rear to where the nose of the craft is, perhaps light years away. In this case, the speed of light as given by 1Lp/1tp is never exceeded, we have merely changed the value of the fundamental Planck length to some very large value via the QZE. In fact, no motion takes place at all. It is a shift from position to superposition to position, all via the QZE. The greatest misunderstanding of the Alcubierre Space-time Manifold is that it is a thing of motion; it is a construct of superposition, only, time, motion, and velocity never occur in the common sense. To date, no one has put forth any notion regarding what aspect regards velocity, how fast, even what direction it goes (one fellow even suggested it required a conventional rocket 'boost' to resolve the issue), not even Alcubierre himself. Yet, everyone recognizes it will work if manifest.

That is, the Orthodox approach is to consider the manifold as contracted at one end and dilated at the opposite end. From the Orthodox perspective, standing at the center of the manifold, if we are looking at the 'contracted' end, assuming we can contract 2.5 million light years of space-time, we are Superpositioned with the Andromeda galaxy. If, on the other hand, our tail end were to face Andromeda and dilate to 2.5 million light years, our tail would be Superpositioned with the Andromeda galaxy. Thus, the Alcubierre Space-time Manifold is not a thing of velocity; it is a thing of superposition, *only*. *No motion takes place.*

Perhaps you can understand my departure from the Orthodox view.

However, if we were to contract 2.5 million light years of space down to an inch, regardless of being Superpositioned, the fundamental Planck length means that we have to cross 2.365×10^{57} Planck intervals of length to cross that inch. If we have not altered the value of the Planck length, then we have a very, very long journey ahead to gain an inch. The Orthodox proposal that the Planck length is fundamental and immutable, *and* that we are going to get from A to B by *contracting* space-time ahead of us is flawed. Remember, my meter stick in my local set of conditions remains quantized to my local conditions. So as I try and cross that 'contracted space,' each Lp I cross = 1Lp, thus, the trip to Andromeda is 2.365×10^{57} Planck intervals, 2.5 million light-years. There is no way around it by 'contracting space.'

This is also true for the photon. However, I do not think the average reader is ready to comprehend the superpositioned nature of the photon yet. Photons also do not 'travel' in the conventional sense, just as we explained earlier that a photon does not 'contract' a billion light-years of space-time to get from A to B, exhausting more energy than the entire cosmos consists of in order to do so (10^{137}). (The visible cosmos weighs about 10^{53} Kg, or 10^{69} joules of $E=mc^2$ energy of ordinary matter, which makes up 5% of the total mass-energy, making the total mass-energy just a bit larger at about 10^{70} joules).

The Alcubierre Space-time Manifold, as we shall go into detail, is a superpositioning manifold. Extend the nose to the target at any distance, and snap the rear of the ship up to meet the nose. This is accomplished by length dilation, space-time inversion (like the LIGO image), as shown and proven occurs in nature by the LIGO data shown above (in the modeling data that matches the captured data). This also negates the need for 'negative energy' or 'exotic matter,' because the negative sign in front of the entire equation is dropped, as we shall see.

This is infinitely more palpable than pulling the distant star to us. We are not 'contracting space,' the classical depiction requires pulling the entire massive gravity well of the distant star as well if we are to contract space to approach the star. The ridiculous proposed energy requirements result from 'squeezing' all of the mass energy of the QED vacuum between us and the distant star into an artificial volume of space-time that cannot possibly exist. That is, contracting all of the mass-energy of space-time, not to mention the QED vacuum energy of that space-time, 10^{113} joules/cubic meter (which would require 10^{143} joules of energy to get to Andromeda), and the mass-energy of a distant star system over a distance of many light years down to a few million kilometers, for instance, is a ludicrous approach. This idea came from science fiction writers. Worse yet, at some point someone took a sheet of paper and bent it in half demonstrating a 'fold' in space. These all become urban myth that our kids grow up with, go to college, get PhDs, and teach other students...

There is also no time dilation factor, because the change in both Lp and tp are 'real,' not *observed effects. We are not actually bypassing Special Relativity, and there is no provision for these limitations in General Relativity. In either case, there is no upper limit for velocity if we alter Lp and tp such that the ratio 1Lp/1tp remains constant.*

The only limitation on velocity will be, in this prototype engine design, which uses electron-positron emission and detection in high populations (referred to as luminosity), and detection and measurement of their spin states by a Stern-Gerlach type arrangement:

- Particle production luminosity
- Particle production velocity
- Particle detection and measurement rate
- Computation power and speed

QUANTIZED MOTION

New Laws of Motion on a Planck Scale

We still need to proceed with background information before detailing the actual engine specifics. Please read carefully and be patient, thank you. If you do not understand these background issues, understanding the engine design will remain a mystery of unorthodox thinking.

The phenomenon of length contraction and/or dilation due to velocity has interesting implications on a Planck scale. We will discuss what some possibilities are. Unfortunately, the problem does not go away when we scale up to the macroscopic. For instance, if we are to look at any given velocity, in this case 0.5c, we have to observe the following rules: (considering that 0.5c is merely $1L_p/2t_p$ and therefore seemingly unambiguously quantized).

Here I must reiterate: First we address the issue that some theorists argue, albeit without evidence or compelling reason, that the Planck Length, L_p is not the fundamental length allowable in this cosmos. So we look again at the equation for L_p, whose derivation is not debated:

$$L_p = \sqrt{\frac{hG}{2\pi c^3}}$$

In the denominator we have 2 pi c, all rock solid constants, albeit pi is a bit fuzzy. In the numerator we have G, which is considered solid, and h. L_p is entirely dependent upon h, the fundamental quanta of this cosmos. If, then ,else: If L_p is not the fundamental length of this cosmos, then h is not the fundamental quanta of energy in this cosmos; else, L_p is the fundamental length in this cosmos and h is the fundamental quanta of energy in this cosmos.

To date, the counter arguments without evidence or reason have been to make way for improvable and unworthy hypotheses, taught in Grad programs as canon to a new Orthodoxy of pet ideas.

Planck Length, L_p, is the smallest allowable slice of space: $1.616199(97) \times 10^{-35}$ m.

Planck time, t_p, is the smallest allowable slice of time: $5.39106(32) \times 10^{-44}$ s

In addition, as it turns out, the speed of light is defined as $1L_p/1t_p$

At v = 0.5c, we are faced with (we review this)

1. go ½ L_p in 1 t_p

That is not possible because this requires a structure finer than L_p (a Planck length) will allow:

Figure 1

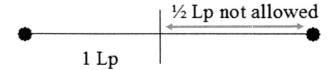

2. Or, go 1Lp in 2 t_p:

    Since proceeding at >1Lp/ t_p is exceeding light speed, this is forbidden.

    Since going < 1Lp is forbidden because it requires a structure finer than Lp will allow.

We are faced with motion taking on the following characteristic:

Go at v=c for 1 t_p, stop for 1t_p, go at v=c for 1 Lp, stop for 1 t_p; etc.,

interval	v	behavior	t'	L'	m'
tp1	c	go	infinite	0	infinite
tp2	0	stop	t0	L0	m0
tp3	c	go	infinite	0	infinite
tp4	0	stop	t0	L0	m0
tp5	c	go	infinite	0	infinite
tp6	0	stop	t0	L0	m0
tp7	c	go	infinite	0	infinite
tp8	0	stop	t0	L0	m0
tp9	c	go	infinite	0	infinite
tp10	0	stop	t0	L0	m0

NOTE: I used the historic reference to length contraction here so as not to confuse the Orthodoxy any further than the argument already will suggest. I do not want this argument to be considered an artifact of inverting Lorentz equation. In this case, inverting the equation is irrelevant.

And so on. Where v is velocity, t' is relativistic time dilation, L' is relativistic length *contraction*, and m' is relativistic mass increase. The values t0, L0, and m0 represent the conditions 'at rest,' or stationary. In addition, t_p represents the Planck interval of time, in sequence; in this case, ten Planck intervals of time and five Planck intervals of length are traversed to demonstrate motion on a quantum scale.

It is not possible to travel one Planck length in distance at any 'velocity' less than *c*, because that requires splitting a Planck unit of length and/or time into *slices* smaller than space-time will allow. It is not possible to travel one Planck unit of distance at any 'velocity' greater than *c*, because that violates Special Relativity. Therefore, on a Planck scale, only the velocities zero and *c* are

possible. Motion on a Planck scale is thus quantized into jumps, or leaps, alternating between zero and c.

Furthermore, since each Planck volume of space-time is isolated from each and every other Planck volume of space-time, this phenomenon of quantized motion occurs separately for each Planck volume of space-time with no apparent means of coordination between Planck volumes of space-time. There is thus some unifying factor involved in this phenomenon. Just as there is some unifying factor that provides a seeming continuity of the progression of Planck intervals of time (t_p) or seeming continuity of space (Lp), mass-energy, the forces of nature, and so on, referred to as the 'Planck Flow.'

The continuity of Planck intervals will be discussed later.

This, stop-go motion characteristic is puzzling enough. However, there does not seem to be any reasonable work around in quantized space-time. This represents a new law for motion on a quantum scale, and it is unambiguously quantized. Note that although m' becomes infinite for a period of $1t_p$, t' has expanded to infinity, thus, the *information* regarding the fact that m' is infinite is 'going nowhere,' frozen in time, is not realized out to any distance, not even out to 1Lp. All motion on any macroscopic scale must conform to this pattern of motion, albeit, the macroscopic is merely a hidden plethora of Planck volumes of space-time (roughly 10^{-105} meters^3). *Information* regarding infinite mass confined to one Planck interval of space-time is perfectly valid. It is only when this *information* extends beyond one Planck interval of space-time that infinite mass becomes a problem. This is in agreement to *information entropy* in Holographic Theory.

So, we consider riding a bicycle. A single neutron contains approximately 10^{60} Planck volumes, each completely isolated one from another such that no two share a common 'now.' At a velocity of 20 KpH just the Planck volumes in that one neutron are making these quantized jumps at either v=c or v=0 such that for any one Planck volume, the sum is 20 KpH, and for the entire neutron the sum is 20 Kph, and for me on the bicycle the sum is 20 KpH. No two Planck volumes in the cosmos share a common 'now.' They are all isolated in their own temporal domains.

This fact of complete temporal isolation is an absolute necessity to accept and understand. In fact, in Holography, which asserts that S=A/4, suggesting that the cosmos is in its final resting place of entropy, that is not ΔS, but S, leaves us with a 2-dimensional static universe where time is NOT a valid dimension. In another text, and/or later in this text I correct this issue, as we are obviously not in a final state of entropy, meaning that the Boltzman solution, S=A/4 is not correct.

In fact, it is considered a possibility that as v=c, a Planck volume proceeds forward in time to the next Planck interval of time, it is annihilated and reconstructed or recreated, or whatever synonym you want to use there, at its destination just one Planck interval away. Thus, we don't actually have 'a' Planck interval that existed since the Big Bang. We can say this intuitively since the past *has ceased to exist.* The *information* regarding the past is available via light that has travelled… But photons exist in an infinitely time dilated state, Superpositioned throughout all of space and time from the Big Bang onward to some indefinite future. In this state they preserve *information* regarding the past. Even memory in the physical brain is electrochemical, meaning carried by virtual photons.

Photons are always at v=c, never at v=0. Every other thing in this cosmos alters between v=c and v=0. Thus, the idea that Planck intervals proceed seemingly forward by annihilation and recreation evidenced by the fact that the past has ceased to exist, and the future does not exist, yet, is feasible. What we refer to as the 'Planck Flow' is a flawed definition of a thing that is not actually progressing, as is the cognitive description of it. A Planck interval is standing perfectly still, each 10^{-44} seconds it ceases to exist, and we refer to that as accumulating in the past. The only instant that exists is this 10^{-44} seconds enveloped within a volume of 10^{-105} m^3. There is no flow, no progression, no future to progress to, the future does not exist to progress toward.

As for the mechanism, it should be obvious. If a Planck interval proceeds forward for continuity, then it must do so at v=c. At this point, its mass-energy is infinite, although the *information* of such is not extending beyond that Planck interval, since the duration is zero. If we consider the limit of the *information* with regards to zero time at v=c we get:

$$\lim_{x \to \infty} \frac{n}{x} = 0$$

Meaning that the *information* regarding infinite mass-energy at v=c becomes an infinitesimal, has no effect on its environment at all. That is, what is referred to as the Planck Flow is quantized, each frame has a life span of 10^{-44} seconds, the progression to the next 'frame' occurs at v=c, such that the duration of the travel time for such *information* equals zero, a 'quantum jump.' The former Planck interval has ceased to exist. Only photons communicate the *information* of its existence to the exact present. Photons do not experience or exist in the Planck Flow of time.

The error in the Boltzman solution, S=A/4 in Holographic Theory (hypotheses) is in not understanding this basic principle of time and information with respect to the 'Planck Flow,' which is not actually a flow, but a standing interval. The standing interval ceases to exist every 10^{-44} seconds, at which point the *information* is carried at v=c to the next standing interval. A photon, which does not exist within the Planck Flow of time preserves the *information* of each Planck interval that has ceased to exist indefinitely. In a sense, the Planck interval that has ceased to exist can be thought of as leaving a photon in its wake as it annihilates. This is actually built right into the equation:

$$t_p = \sqrt{\frac{hG}{2\pi c^5}}$$

$$h = \frac{2\pi c^5 t_p^2}{G}$$

Returning the value *h* leaves us with a variety of forms of *information* that can be encoded in a photon.

It is best to think of Holographic Theory in its current primitive form as a shell, a Swarzschild surface just one Planck interval thick, expanding outward from the Big Bang as a foci and the only temporal marker that exists. As the surface expands, being one Planck interval thick, the former *sphere**, like an onion, is annihilated, and a new razor thin layer is born. Thus, the surface is always 2-dimensional, with the foci being the Big Bang. However, the Swarzschild surface is expanding, albeit remaining exactly one Planck interval thick. It is incorrect to think of it as a hollow sphere, it has no interior. All of the Planck intervals of space and of time have ceased to exist in what we would think of as the interior.

*Sphere is a difficult term to use here as it designates a dimensionality greater than 2. However, as I stated, S=A/4 is an incorrect result.

'Observed' Length Contraction on a Planck Scale

Lp', tp', and G'

More puzzling is what happens to the observed length Lp', due to relativistic length 'contraction.' Again, when I measure someone else's value Lp under relativistic conditions, I measure against my locally quantized meter stick, and the values for Lp differ. As with mp and tp, this is not merely an observed effect, a real change has taken place, albeit, each observer in their own frame of reference measures no change of their own condition, leaving their meter stick quantized to their local conditions, again.

Historically, this condition was thought of as an observed effect because of the seeming paradox it raised, a meter stick having a different value here than it does over there, with no consistency then throughout the cosmos. However, there is no consistency to time, that is a given, there is no moment we can finger as a common *now* to any two Planck intervals of time anywhere throughout the cosmos, not even common to two brain cells, nor is there any common clock we can figure as governing any such common time for any two Planck intervals of time throughout the cosmos, from the Big Bang until the present. There are currently roughly 10^{184} Planck intervals of time (regarding them as Planck domains volume filling) in the visible cosmos, and no two share the same *now, the same clock,* and it is not a paradox. In fact, if and when any two do seem to share the same clock, we call that *quantum entanglement, or Superpositioned,* and *that* is a paradox. Time and distance (length) are bound by the fundamental relationship of the invariance:

$$c = \frac{1Lp}{1tp}$$

In order for *c* to remain truly invariant, Lp must be as elusive as tp, with no common meter stick, and this turns out to be the case as we measure most distant objects as possessing some degree of redshift or blueshift. Either objects are in proper motion or in the Hubble flow, relative to us observers.

Using the classic Orthodox approach of *observed* length contraction on a Planck scale: (for lack of confusion regarding this being an artifact of inverting the equation)

$$Lp' = Lp\sqrt{1 - (\frac{v}{c})^2}$$

At a velocity of 0.5c, Lp' is roughly 0.866. This value, using an arbitrary precision calculator, goes out to at least 1000 decimal places. In addition, of course, a value of 0.866Lp is not allowed because it requires a structure finer than Lp will allow.

0.8660254037 8443864676 3723170752 9361834714 0262690519 0314027903 4897259665 0845440001
8540573093 3786242878 3781307070 7703351514 9849725474 9947623940 5827756047 1868242640
4661595115 2791033987 4100505423 3746163250 7656171633 4516614433 2533612733 4460918985
6135235658 3018393079 4009524993 2686899296 9473382517 3753288025 3783091740 6480305047
3801093595 1625415729 1476197991 6498894912 2541443572 3191645867 3612081992 2939276988

3397903190 9176833055 4215868904 4718915805 1044152762 4508350117 6035557214 4347995478 1828985435 8424903644 9746648242 1415103932 0430199436 9348768791 1586589156 9799649150 3919351438 5269566847 8165605185 3632009624 5533841155 9964418782 0570711008 3713760511 8649713541 5529949229 7379938321 4444889807 3918979195 1144274264 5178801692 6404032190 9861723305 2984486143 6432632076 9113323492 1001059774 2077639220 5906432672 5351759582 5008344647 2077404230 3563857199 9881463417 3147887191 8094755506 3574319373 4882729912 2589427548 7689506940 3324809559 8111147855 5277621461 8615960988 6913128081 5734421016 4268583414 6932480595 8524869418 1977479691…etc.

The argument, this is an observed effect, is exactly the point. Regardless of what is actually happening aboard our spaceship at $v = 0.5c$, I, the observer, who am presumably stationary, have a meter stick which has no choice but to be quantized according to Lp *in my stationary frame of reference*. There is no possible way for me to 'observe' a value other than an integer value of Lp as Lp is measured in my stationary frame of reference, because in my reference, Lp is still Lp. *It is noteworthy that my arbitrary precision calculator, surpassing 1000 decimal places, also surpasses the order of magnitude of the Planck scale.*

Common discussions regarding Lorentz transformations and such, regard macroscopic scales, light-years, where little is discussed regarding such transformations on a quantum scale. In general, such discussions are limited to math that is highly complex and targeted at phenomenon of an unusual nature. In this discussion, we are limiting our argument to the simple, velocity, the transformation characteristic, and the Planck scale that *must* scale up to a cosmological scale.

We could say that 0.866 out to at least 1000 decimal places becomes truncated at some point, say 866/1000, such that Lp will be 'observed' by me as taking on *near but arbitrarily inaccurate* unit values of Lp. However, the point of truncation becomes arbitrary, and also requires that length contraction (and/or dilation) due to velocity cannot be observed at, say, 277Lp (which is less than 866Lp) distance, because this would then render a non-integer value of the arbitrarily chosen 866Lp at this arbitrarily agreed upon degree of precision.

Since I am stuck with a quantized meter stick, and have no choice but to 'observe' integer values of Lp, this requires either a change in precision, an arbitrary truncation of our 0.866… value, or otherwise, motion MUST occur at intervals of no less than 1000Lp, assuming we are satisfied with a precision of 866/1000, which is arbitrary. However, as shown above, the value 0.866 is 3 decimals of greater than 1000 decimal places precision, far surpassing Planck's order of magnitude. We are truncating for no apparent reason other than confusion or refusal to face the complexity of the issue at hand.

The greater the precision, the greater 'leaps' of distance, motion in bundles of 1000Lp, for instance, would be required. If I go out to 10 decimal places (0.8660254037Lp), then in order to get integer values of nLp/xt_p, I now have to make jumps of Lp that are becoming macroscopic values (on the order of microns).

If we do not make 'jumps' in bundles of 1000Lp, then our precision, our actual measurement, is changing with each Lp distance traveled.

If you do the math for a macroscopic object, the problem persists. It does not go away and fade into the background of measurable precision. In fact, the act of *acceleration* that requires a constantly changing ratio of nLp/xtp^2 becomes so bizarre that it is apparent we have missed something in some definition or equation somewhere.

If we take into account Wheeler's Space-time Foam on a quantum scale, [John Archibald Wheeler with Kenneth Ford. *Geons, Black Holes, and Quantum Foam.* 1995 ISBN 0-393-04642-7.] we might conclude that as a part of this foamy characteristic of space-time on a quantum scale is the motion of a macroscopic object progressing forward in this go-stop-go fashion described above at v=c and v=0. In today's vernacular we might say that the object were moving as though pixelated, and as we back out from the quantum to the macroscopic we no longer see the pixelated but a 'normal' progression of a macroscopic object. However, there can be no 'pixelation' on a quantum scale, as I will describe later on, because of the foamy characteristic of space-time on a quantum scale. In fact, there can be no shape, again, this will be described.

It is vital to keep in mind that the person aboard our spaceship measures no difference in condition with respect to Lp. The length contraction (and/or dilation) is limited to me, the 'observer, and I am stationary, and I have no choice but to measure and 'observe' the speeding object with my quantized meter stick; which is quantized to my stationary frame of reference. However, the person aboard our spaceship has a meter stick quantized to his local frame of reference and when he goes to measure me will measure a value different from mine, yet quantized. Two values for Lp, both real, but differ.

In no uncertain terms, there can be no 'smooth' transitions or non-quantized values of Lp measured as a result of length contraction and/or dilation because my meter stick by which I measure such phenomenon is quantized to my stationary frame of reference, and such is also the case for the speeding traveler measuring me. Both Special and General Relativistic changes are therefore not 'smooth' transitions or transformations but quantized according to the observer's locally quantized meter stick, again, such is also the case for the speeding traveler or he/she in a gravity well.

If the traveler used me, the stationary observer as a reference, the traveler would be using his quantized meter stick in his frame of reference as the reference. In this case, the speeding traveler would face the same problem of using his locally quantized meter stick to measure relativistic changes in his observations of me. The result will therefore be in integer values of nLp and no fractional value of Lp is possible for either the traveler or stationary observer but neither result agrees with the other. This entire argument translates to General Relativity equally except we place the two observers at different points anywhere within a gravity well.

Although this subject may seem exhaustive, it is vital to the design of Alcubierre's Space-time Manifold. The entire manifold must be quantized or it will not function. Even from the Orthodox point of view, Lp along the leading edge will certainly not agree with Lp along the tailing edge of the manifold. That is, the manifold has severely different quantizations just from front to back, both in space and time. The very scaffold of space-time, the unitary Planck interval is magnitudes of order different on one end from the other, a problem no one has considered to date. In fact, up to this point, the Planck unit has been considered fundamental and immutable. However, this cannot be the case, as an axiom, as we have seen via relativistics.

In fact, the manifold's side, which has a slope, has a sliding value of nLp gradation in space-time quantization that if not faced head on as the most problematic feature of the engine design will result in failure.

Furthermore, a fundamental understanding of Lorentz transformations, with respect to observer/observed quantizations is imperative. Oddly, these issues have never been discussed with respect to Alcubierre's manifold before, when it is obvious when pointed out that it is impossible to differentiate contraction from dilation without a thorough and exhaustive analysis of quantizing this approach.

As we will investigate a bit later on, transcendental values such as pi, non-integer values of Planck units, and so on, will cause the manifold to fail because the manifold must be precise down to a Planck scale. I have therefore rendered a complete section of this engine design to correcting Alcubierre's Space-time Manifold down to a Quantized scale and out to cosmological distances. Again, the engine functions by extending the value Lp out to cosmological distances, then snapping the rear of the vessel up to meet the nose of the ship. *Velocity* may not be the right terminology to use in this case. Billions of light-years can be traversed *instantaneously,* without actual motion.

Again, the traveler's 'meter stick' is quantized. However, the quantizations do not agree with one another. In order for the two quantizations to not agree with one another, a 'real' change must take place in some frame of reference, or both frames of reference, and they *must be integer values of one another*. Ultimately, if we deconstruct the entire scenario on a Planck scale out to cosmological distances, it would then become obvious that each isolated Planck volume of space-time has its own unique frame of reference and there is no universal Planck Length or Planck unit of time. Recessional Velocity demands Special Relativistic application, and space-time expansion, be it accelerating or decelerating, demands General Relativistic application. We will examine this phenomenon shortly.

What did you think space and/or time dilation is? Is it a squeezing of Planck intervals together tighter from the distant observer's perspective as though there were some play between them to work with? Is it the loosening of Planck intervals from the traveler's perspective? If the Planck intervals do not change under Relativistic Conditions than these are the only options. That being said, an alteration in the Planck interval, in Planck's constant, and as such all of the subsequent dependent 'constants' of Planck's constant, G, for instance, MUST change under Relativistic conditions. Otherwise, we are faced with the idea that space-time 'squeezes' Planck intervals of space and time together (from the stationary observer's preferential perspective) or softens them up (from the traveler's preferential perspective), which is not possible because the idea that there is any 'play' between Planck intervals in order to accomplish this is ludicrous and improvable. I honestly do not think anyone who has given it a thought has followed through with that thought for fear of opening a can of worms. Either Planck intervals are loosely packed or the otherwise immutable Planck interval is mutable, there is no third option.

Under Relativistic conditions, Lp, t_p, change, meaning that Planck's constant changes, and/or consequently, G changes, which I will from herein refer to as G'.

What is Planck's constant? Planck's constant is 'the' minimal amount of energy… The terminology splays out into every field of physics and mechanics. We dilate time to infinity such that time comes to a complete stop inside of a sphere in which we are sitting, like a lounge chair inside of a Black Hole. From inside the sphere we watch the passing of the cosmos in a fraction of an instant, the rest of the cosmos has seemed to speed up (temporally) to infinity.

From Larson, Edwards, Calculus, 9th Edition, section 3.5, pg 199, 'Limits at Infinity.'

THEOREM 3.10 LIMITS AT INFINITY

If r is a positive rational number and c is any real number, then

$$\lim_{x \to \infty} \frac{c}{x^r} = 0.$$

Furthermore, if x^r is defined when $x < 0$, then

$$\lim_{x \to -\infty} \frac{c}{x^r} = 0.$$

$$\lim_{x \to \infty} \frac{1}{x} = 0$$

$$\frac{n}{\infty} = 0$$

We can partially resolve this issue by thinking in terms of *xt$_p$* and *nLp*.

$$C = \frac{1L_p}{1t_p}$$

$$V = \frac{nL_p}{xt_p}$$

$$\frac{V}{C} = \frac{nLp/xtp}{1Lp/1tp}$$

The proper quantization of time and length dilation being (Special Relativistic):

$$x't_p = \frac{x_0 t_p}{\sqrt{1 - (\frac{\frac{nLp}{xtp}}{\frac{1Lp}{1tp}})^2}}$$

$$n'L_p = \frac{L_0 t_p}{\sqrt{1 - (\frac{\frac{nLp}{xtp}}{\frac{1Lp}{1tp}})^2}}$$

This is using *length dilation* under Special Relativistic conditions.

THE SHAPE OF SPACE-TIME ON A PLANCK SCALE

We'll discuss G; in a separate argument later on. In essence, you cannot squeeze Lp and t$_p$ at their given values into a Black Hole, the idea is ludicrous. Lp and t$_p$ must change value in order to squeeze down to a Swarzschild Radius. As an example, I use an illustration from one of the first papers presented on the Holographic Principle of Quantum Mechanics:

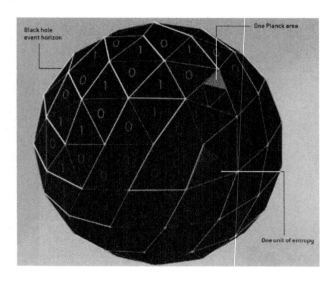

These triangular regions denoting Planck area are an impossibility in normal space-time. A triangle is dependent on values that are not integer values of Planck's constant and therefore not possible in normal space-time:

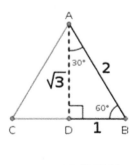

$$DA = \sqrt{DB^2 + BA^2}$$

In this diagram DB=1, BA=2, and DA=√3

√3 is not an integer value of the Planck interval and is therefore a value that cannot exist in normal space-time. As a result, the triangle is a shape that is impossible for the Planck length in normal space-time. Similarly, we find such an equilateral triangle laced with issues:

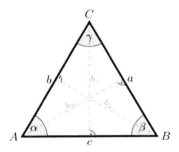

Holography has each Planck area neatly packing down into a 2-dimensional space-filling sphere of Planck triangles each consisting of sharp equilateral triangles of about 10^{-70} meters squared.

Although it is possible to designate a numeric value for the Planck interval of length squared simply as Lp^2, roughly 10^{-70} meters, the area cannot actually take on a physical shape in normal space-time. A sphere for instance has a circumference which is pi times the diameter, pi is a transcendental number with no actual discrete value, certainly not an integer value of the Planck interval.

The volume and surface areas of the sphere are given by:

$$V = \frac{4}{3}\pi r^3$$

$$A = 4\pi r^2$$

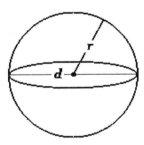

A cube runs into similar problems, although at first thought the cube may seem like a logical choice, the cube is laced with triangles and pyramids all of which defy the integer value of the Planck length.

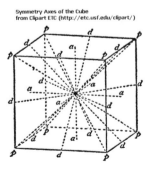

Not only do the hypotenuses crisscrossing the cube create lengths that defy the integer value of the Planck length, but the pyramidal shapes they enclose create volumes that defy the integer value of the Planck volume.

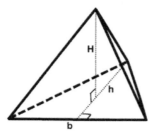

Volume of a Triangular Pyramid = $\dfrac{\text{Area of Base} * \text{Height}}{3}$

Any Planck area, and Planck volume, must be shapeless in normal space-time. It isn't so much that the 'shape' must be shapeless in normal space-time, rather that space-time takes on very different aspects on a quantum scale. ["Quantum Foam". New Scientist. 29 June 2008] On this scale, space-time is not consistent or static, but is extremely dynamic, with fluctuating properties in what one would normally think of as a static, flat, fabric. Virtual particles are appearing and disappearing in great abundance at a phenomenal rate, further upsetting the underlying structure. It is best to think of space-time as being so turbulent on this scale that any concept regarding 'shape' is simply non-sequitur.

That is, a 1-dimensional Planck length can have a 'shape' of a 1-dimensional line, *the line cannot be curved because this requires a substructure finer than Lp will allow:*

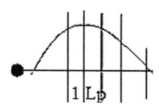

A curved line of one Planck length cannot curve or bend because it requires slicing Lp into pieces finer than Lp will allow. In the diagram above, Lp, in order to curve, has to undergo changes on a scale finer than space-time will permit. This explains my aversion to String Hypotheses as a rule.

Any higher dimensionality (compactification) cannot be represented on a Planck scale because of the inherent problems demonstrated in the above characteristics of shapes involving 2 or more dimensions that require non-integer values of Planck intervals to represent. On a Planck scale, area and volume go shapeless, and conform to the Quantum Foam characteristic of space-time. This means that the system is dynamic, not static. The representation above of a sphere made up of Planck areas of neat triangles is impossible in a Quantum Foam environment, which is highly dynamic, churning at impossible speed; keeping in mind that time is churning on this scale just as much as space. There are no neat, static triangles to be found there. Compactification as described in the various String Hypotheses requires a static environment; they are in fact void of the dimension of time as a rule.

It can be said as a metaphor that part of the foamy characteristic of quantum space-time is the dual between shape and shapelessness. Shape is an impossibility on a quantum scale, yet shape is a demand in 4-dimensional space-time, and yet space-time is a 2-dimensional rendering of a 4-dimensional façade in Holography. All of these characteristics are at war with one another on the quantum scale, where time itself emerges and submerges. Thus the Foam is highly dynamic at the 4-dimensional level and yet static at the 2-dimensional level, giving rise to Kip Thorne's quantum scale Einstein-Rosen Bridges [Thorne, Kip S. (1994). *Black Holes and Time Warps*. W. W. Norton. pp. 494–496. ISBN 0-393-31276-3.]

At this point, one may try to introduce an argument that speculates on higher dimensionalities. However, this is indeed speculative and there is as of yet no evidence of higher dimensionalities other than those that exist in mathematical, speculative models, in which case the term 'Theory' does not apply to any of these cases but they remain hypotheses in the vacuum of any evidence. In any case, the models of higher dimensionality cannot resolve the stated issue because they are encapsulated (compactified) dimensions. That is, they do not extend outward from the Planck interval in order to 'paint' a static figure of a Planck triangular area:

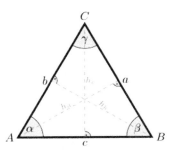

In this diagram, each hypotenuse has to extend outward at some distance as a non-integer value of the Planck length, such as the hypotenuse, whereas the models for higher dimensionalities encapsulate these higher dimensions *within* the Planck length. Therefore, calling on higher dimensionalities in either case of their actuality does not resolve the issue. Furthermore, as stated, the Quantum Foam will simply not allow this static triangle to exist, given that each leg of the triangle is to be one static Planck length. The effect of a highly dynamic Quantum Foam would be much more severe on any set of encapsulated dimensions within each leg of the triangle.

For example, we imagine a set of compactified dimensions like those within a garden hose. A higher dimensional line of sight point at B would completely lose sight of a higher dimensional point at A altogether as the leg of the triangle bends to conform to the Quantum Foam. Where line of site is a given characteristic of an encapsulated dimension, such a characteristic would be lost. Just as line of sight is a characteristic on a macroscopic scale to us, losing such line of sight due to extreme space-time curvature would present a problem if not a paradox. This is the *opposite* of gravitational lensing, this is the complete loss of line of sight altogether as a result of the turbulent nature of the Quantum Foam.

Thus, when we have reached the point where we are close to the Swarzschild radius, keeping in mind that due to gravimetric time dilation the approach to the Swarzschild radius is asymptotic and can never actually be reached but is forever approached, we have squeezed all of our Planck areas down to something like the depiction of our Planck triangles, albeit not tringles but shapeless. Furthermore, the Quantum Foam has settled down from a raging storm to a quiet sea due to extreme time dilation, and the region has become nearly static. t' Hooft's query regarding the dynamics of the *information* falling into the system is partially resolved in that respect, at least, albeit I do not think he was thinking of the Quantum Foam when he posed the question. However, the *information* contributes to the foam and must be taken into consideration. The *information* contains dynamic quality to it that in turn churns the foam. In any case, gravimetric time dilation has caused the system to become nearly static, but the issue regarding the Planck areas remains.

Again, contrary to String Hypothesis and so on, it is not possible to project *information* from a compactified dimension of less than 1Lp (1t_p) because that would require projecting *information* in packets of less than Planck's constant, h. Since h represents the smallest slice of *information* allowable in normal space-time, there is no mechanism, known or hypothetical, for taking any value smaller than h and making it somehow 'visible *information*' in normal space-time. Thus, the idea that particle physics is an emanation from compactified dimensions is quite impossible.

When we reduce the Planck value to purely Temporal Dynamics, we only have three choices to work with:

$$\{-1, 0, +1\}$$

Because of our previously discussed issue with trying to 'bend' a String of one Planck length:

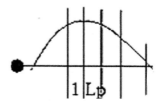

We are not going to get this resolution because of the Planck limit, only $\{+1, 0, -1\}$. The diagram above clearly shows an attempt to slice one Planck length into slices finer than normal space-time will allow.

We can eliminate zero, as this seems by definition to have no meaning in normal space-time, and reduce the possibilities to:

$$\{+1, -1\}$$

Furthermore, the concept of a wave function in the form of a 1-dimensional string of one Planck length doesn't work:

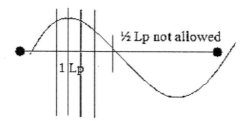

This requires taking slices of the Planck length that are not allowed in normal space-time. Encapsulating dimensions *within* the Planck length does not resolve the issue as we discussed. The wave function has to exhibit wave characteristics beyond the confines of the encapsulated domain of the Planck length in order to shed its wave-like character beyond the Planck barrier or domain. That is, in order for the one Planck length wave function to carry its wave-like character outward in normal space-time it has to exhibit wave-like character on a Planck scale. Wave-like character contained or restrained within the confines, encapsulated within the Planck domain will yield no effect in normal space-time. Again, on a Planck scale, the Planck can only exhibit the following characteristics:

$$\{+1,-1\}$$

The *information* is binary. If it were possible to separate space and time, +1 could represent for instance, the distance A to B, and -1 the distance B to A. Temporally, in common time we would make the same statement. The difference is, in Temporal Mechanics, once you leave point A it no longer exists, and B does not exist *yet*. Thus the distance, A to B becomes a floating variable with no distinct endpoints because the starting point has ceased to exist and the endpoint has not yet come into being. Therefore, the values {+1,-1} in Temporal Mechanics become pointers, only, whereas in spatial mechanics they can be meter sticks with substance and permanency. That is, in Temporal Mechanics +1 can represent forward time, and -1 can represent backward time, *on a macroscopic scale.*

On a Planck scale, that issue is an entire text [Temporal Mechanics 101, Dr. William Joseph Bray ISBN 978-1539410959]

Thus, we are left with several dilemmas. A wave-like characteristic originally argued as a single Planck length resonating like a string, but this is not possible, as this requires parsing our 1-dimensional one Planck length string into subunits smaller than space-time will allow. A static triangle of one Planck area that contains at the very least a hypotenuse of a non-integer value of the Planck length and therefore the shape is impossible, and indeed any 2 or 3-dimensional shape is impossible on a Planck scale for the same reason of being wrought with non-integer Planck values. A static 2-dimensional shape in a raging turbulent Quantum Foam that will not allow a static shape to hold any form on a Planck scale. And last but not least a turbulent Quantum Foam that has come to a near stop due to gravimetric time dilation altering the characteristics of subspace, which is 'foamy' on a Planck scale by definition, in ways not yet understood because it has never been considered.

This is not to be confused with squeezing the space out from between the particles and Wave Functions that make up particles such as atoms and then nucleons to surpass the density of a Neutron Star and become degenerate matter for a millisecond and collapse into a Black Hole. This is the squeezing of space-time itself, not of the 'stuff' in it. Physicists are all too often complacent dealing with squeezing the space out from between the 'stuff' between particles with disregard to the quantization of space and time and what is happening on that level.

That is, there are as many available units of Lp and tp as there are Planck volumes of space-time, an estimated 10^{60} such volumes in a single nucleon. As a result, I will drag this discussion out to the 'Bag Model' of quark confinement to make my point more clear a bit later on.

This substructure of space-time on a Planck scale is important to understand in order to achieve a working Alcubierre Space-time Manifold capable of extending Lp' out to indefinite length, light-years.

Expressing the Gravitational Constant, G, as a renormalized value, G'

Let's look at time dilation by plugging the time dilation factor back into my Gravitational Constant: G

$$\pm t' = \frac{t^0}{\sqrt{1-(\frac{v}{c})^2}}$$

As an aside, keep in mind that v/c is the same as saying:

$$\frac{nLp/xtp}{1Lp/1tp} ; \frac{v}{c}$$

However, this level of complexity is not required to resolve the problem. It is an important observation to keep in mind, however, because maintaining this ratio as integer values during the process of acceleration is impossible on both a quantum and macroscopic scale unless one progresses in the fashion previously described as quantum leaps in v=c or v=0. There is no smooth non-quantized velocity, nor is there any smooth non-quantized acceleration. Thus, Gravitation, including the gravitational 'constant' cannot be a non-quantized value.

For the following calculations, we will use high precision, because squaring and cubing things produces errors that expand exponentially, so we will use the current level at which these values are reported.

The speed of light: 2.99792458×10^8 m/s

The Gravitational Constant: $6.67384(80) \times 10^{-11}$ m³/Kg s²: Also noting as an aside that this is inverse density per s².

If we take our value for t' at v = 0.5c, which is about 1.1547, and plug it back into our value for the Gravitational Constant G:

$$G' = 6.67384(80) \times 10^{-11} \text{ m}^3/\text{Kg (s)}^2$$

Substitute t' for s

$$G' = 6.67384(80) \times 10^{-11} \text{ m}^3/\text{Kg (t')}^2$$

$$G' = 6.67384(80) \times 10^{-11} \text{ m}^3/\text{Kg (1.1547)}^2$$

$$G' = 5.005386 \times 10^{-11} \text{ m}^3/\text{Kg s}^2$$

A subsequent 'renormalization' is not necessary for this argument. Later, we will see that G' is a fractal of infinite self-similarity.

We then re-introduce this new value for G' back into our equation for Lp:

$$Lp' = \sqrt{\frac{hG'}{2\pi c^3}}$$

And the numeric result, when we substitute our newly derived value for G' at v = 0.5c is Lp' = 1.398…x10^{-35} meters depending on your truncation. The new value for Lp', which we will designate Lp$_{G t'}$ then looking at the ratio:

Lp$_{Gt'}$/Lp = 0.866….

That is

$$\frac{Lp\prime}{Lp} = \frac{\sqrt{\frac{hG\prime}{2\pi c^3}}}{\sqrt{\frac{hG}{2\pi c^3}}} = 0.866\ldots$$

(It is actually the same out to 1000 decimal places). Which is the same as Lp' at v = 0.5c according to *the classical Orthodox representation of the equation, so as not to argue an artifact of inverting the equation:*

$$Lp' = Lp\sqrt{1 - (\frac{v}{c})^2}$$

Since G is not derived from t' (in Newtonian mechanics) and vice-versa, so as to reach Lp', it is not a mathematical artifact.

This value in this example was achieved by replacing the Newton in s^2 with t'^{2} as shown above. This shows, conclusively, equivalency between the values Lp', t$_p$', and G'. G' is a product of Relativistic conditions, and is also quantized.

The actual observed value for Lp', as a stationary observer, with a quantized meter stick, has changed as a direct result of t' having changed of the observed, speeding object, changing the otherwise unchangeable Gravitational Constant, G, affecting the actual value for the Planck length, Lp – *for the observer*, regardless of my quantized meter stick.

Note here that I am using the classical definition for Lorentz equation as a demonstration that this phenomenon is not an artifact of inversion of the equation. The net result is the same. I do not want to further confuse the issue of the Orthodox views by flipping equations, regardless of them being upside-down. The Orthodox view will suffice for the sake of argument for now.

The result here is not evidence of length contraction.

When we extend this argument to tp' according to:

$$tp' = \sqrt{\frac{hG'}{2\pi c^5}}$$

At o.5c tp' is 4.6688 x 10^{-44} s and the ratio tp'/tp is 0.866…

The ratios, Lp'/Lp and tp'/tp are identical. The invariance c=1Lp/1tp is preserved. They are not reciprocals of one another. This validates:

Given that c=1Lp/1tp

Again because:

$$c = \frac{L_0\sqrt{1-(\frac{v}{c})^2}}{\frac{t_0}{\sqrt{1-(\frac{v}{c})^2}}}$$

$$c = \frac{L_0}{t_0}\left(1-\left(\frac{v}{c}\right)^2\right)$$

We can then say that if

$$l' = l_0\sqrt{1-(\frac{v}{c})^2}$$

Then

$$c = \frac{L_0}{t_0}\left(1-\left(\frac{v}{c}\right)^2\right)$$

When v=c; c=0

The Orthodox representation of *length* contraction under Relativistic conditions cannot be correct because when v=c; c MUST = 0, which is obviously incorrect. Therefore:

$$t' = \frac{t_0}{\sqrt{(1-(\frac{v}{c})^2}} \qquad tp' = \frac{tp_0}{\sqrt{(1-(\frac{v}{c})^2}}$$

$$l' = \frac{l_0}{\sqrt{(1-(\frac{v}{c})^2}} \qquad Lp' = \frac{Lp_0}{\sqrt{(1-(\frac{v}{c})^2}}$$

And also that G is a variable, G', not an immutable constant. The variable G' is given by:

G' = 6.67384(80)×10⁻¹¹ m³/Kg (t')²

The reason m³ is held constant and not treated as Lp'³ but t'² is treated as a variable is that m³ is not changing. G is expressed as an acceleration, a value that is changing over time. Thus t' is a variable, the only variable in the expression that has any *immediate* effect on the outcome of the system. Where we could argue that the meter changes as Lp changes, this level of renormalization isn't necessary for G'.

We then go back to our scenario where we are measuring a photon, as viewed head on, with a redshift of '2'. Our quantized value Lp is used to directly measure the photon as it passes over our head, and we require twice as many Lp to make that measurement. In the case of tp', we measure the unit length per unit time (Lp/tp) according to our local value tp in order to determine the length. Again, the above stated relationships are preserved.

This brings us back to our previous discussion we left hanging regarding h as a variable that seemed to be expanding toward infinity as G' is shrinking away toward zero. G' is a dynamic 2-dimensional acceleration, it has its own meter sticks by which to gauge itself in 2-dimensions; h is a 1-dimensional meter stick with no apparent universal meter stick by which to gauge its value. That is, in the case of h, we have no universal meter stick by which to measure it, as is the case with the observer in each frame of reference, no change can be monitored in any local frame of reference, therefore no apparent change takes place, and the hidden variable, h, remains intangible. It isn't until we reach the extremes of infinity and infinitesimals that we see h produce an effect in all frames of reference, this given that the cosmos is in itself finite. The value, h, is in fact the 'constant' by which we meter the value of all of the other variables and constants. Thus, we observe changes in these alternate values metered by h by holding it constant.

These relationships are fundamental in understanding the working principle of the FTL engine based on Alcubierre's space-time manifold, as will be shown to require zero energy to produce other than the detection system.

*In one sense, if we travel from Earth to Alpha Centauri in one Planck unit of length that has been extended to 4.4 light-years, semantically, we can regard space as being 'contracted,' 4.4 light-years down to one Planck length. That is the observed effect. The method for achieving this is via *the real cause, extending the Planck unit of length out to four light years*. In this paper, we need to keep the observed and real effects unambiguous.

Borrowing a small fraction of the many definitions for Non-Locality, if the travel time from the photon's perspective is zero, the distance is zero (semantically, we say 'contracted'), but then the photon must be non-localized such that it is both 'touching' its starting point and destination, and thus we say it is *superpositioned*. Although this may seem intangible, this is exactly the world of the photon. To the photon, all points in the cosmos are at zero distance and travel time. In effect, the photon is Superpositioned throughout the cosmos and within its own frame of reference, never travels any time or distance to any local. But from the photon's perspective, as odd as it may seem, every corner of the cosmos is Superpositioned into one state, one local. Again, Schrodinger's equations show the photon as Superpositioned prior to wave function collapse. To the photon, the cosmos is a superposition of every point in the universe collapsed into one quantum state that exists as an infinitesimal.

Now we break out those quantized meter sticks. Each is now quantized to 'zero.' The photon is Superpositioned throughout the cosmos, (in agreement with Schrodinger's equations, Von Neumann, etc.) and every corner of the cosmos (from the photon's frame of reference) is Superpositioned into one state. Zero is a perfectly valid quantization. It would seem we are left with only two irrefutable universally agreed upon quantizations:

$$\lim_{x \to \infty} \frac{n}{x} = 0$$

And

$$\lim_{x \to \infty} \frac{x}{n} = \infty$$

I will therefore clarify unambiguously: It is not possible to violate Special Relativity, even by the proposed mechanism of the 'upside-down-backward' classic depiction of the Alcubierre manifold, namely because it is upside-down and backward, and General Relativity arguments fail because the system is non-localized. The only way to seemingly exceed the velocity of light is to change the fundamental values of Lp. By extending Lp out to great distances, say, one light-year, one can travel apparently without limitation to regard to velocity by traveling 1Lp/1tp. In this example, we have traveled one light year in 10^{-44} seconds without exceeding the speed of light.

This evolution of events results in Lp' and tp' expanding out to infinity as both (+/-)t', fulfilling all of the characteristics of non-locality both in the presence or absence of quantum entanglement.

We continue the argument based on the reference of the speeding traveler. If I am the speeding traveler, then I am measuring no change in my own condition. Since all values in my frame of reference are quantized, then again, the Planck values, time, space, and so on, remain quantized.

Then we are left with the fact that there is no frame of reference where a non-quantized value can be measured. The quantization of time dilation, *change in length,* and so on remain quantized in all frames of reference, given that any measurement must be made via my local, quantized value for Lp and tp, or from another preferential frame of reference wherein their meter sticks are also quantized to their local values under their local conditions. That is, everything remains quantized to values within their local frames of reference, albeit the quantizations differ from one another, the values are real, not observed, just as the temporal change in the rate on a clock under relativistic conditions is real, not just an observed effect.

The implications are that on a cosmological scale, in the Hubble flow, no two locations at distant points in the cosmos measure the same quantized value of one another. Zooming in to a galactic cluster under gravitational forces, again, the Ricci curvature of the cluster will yield a result of differing values of quantization at galactic distances, and so on down to stellar distances. This issue persists, as discussed, all the way down to 1 Lp distance where information must be transferred at v=c or v=0 and as such are set to the limits shown in the equations above. As such, there are a near infinite number of meter sticks out there in the cosmos, no two agree. The system can said to be chaotic. The only chaotic attractor in the equation is that in my local frame of reference it does not matter what all of the other near infinite number of meter sticks say, only that when their light reaches my meter stick they do so in a predictable fashion according to distance and all of the other Orthodox 'stuff;' Hubble flow, proper motion, comoving distance, and so on. We will look at the Hubble flow in detail shortly.

Since we have exhausted the possibilities in all frames of reference the inescapable conclusion is that the *real values, Lp and tp have undergone a real change. Just as space and time were once considered fixed and immutable, our current philosophy of 'fixed and immutable' constants possess gaps, many of which have led to additional 'paradoxes.'*

This variability of Lp', tp', and G' would seem to make one consider that the speed of light itself is variable, at first glance. I have seen more than one paper dedicated to examining this possibility. However, the variability of Lp', tp', and G' resolve the issue, and *c* remains constant in all frames of reference. We can consider it that God has left us with just one universal and eternal meter stick, 'c.'

The value, Lp according to my meter stick *should* differ from the value Lp' that I measure for a speeding traveler under Special Relativistic conditions. *However,* I cannot measure a value of Lp' that is different from my own because my meter stick is quantized to my local environment. This is why the recalibration of a GPS satellite has to be reset manually and cannot be reverse engineered to run faster or slower to auto-compensate. This is why some people get confused and believe that the effect of observing a change in length is only *observed and not real.* All observations, detections, and measurements are reliant on my local quantized meter stick. On a macroscopic scale, things appear different, but on a quantum scale, they *must be* the same because of local quantization. We then have to *balance the local values of the quanta per quanta, and the must be in integer values:*

What's more, during acceleration or climbing in or out of a gravity well, we have to quantify a constantly changing balance of quanta per quanta in integer values:

A GPS satellite in orbit experiences quanta building up on one side as a result of ticking faster (lightening the load of quanta via General Relativity in high orbit) and ticking slower (heavying the load of quanta via Special Relativity via velocity in orbit). Since there is no fix at the quantum level at this time, the quanta have to be shaved off one side or the other manually by simply resetting the internal clocks to match those on the ground. Then the process begins again for another few hours. Again, there is no autocorrecting for this, no running a bit faster or a bit slower according to some mathematically predicted value, it has to be done manually. If it is not done in 24 hours your GPS will be off by 7 miles (if I recall correctly).

In all frames of reference, the changes in Lp and tp must be real.

This is not off subject. The engine design takes advantage of these real changes in the values Lp and tp. As we will examine later, the principle of the engine is to extend the value Lp out to some macroscopic distance. Since we have already examined that quantized motion can only occur at v=0 or v=c, the engine will be making these 'jumps' at v=c, the only possible value the engine can move, or otherwise be stationary. However, the value Lp will have increased by many orders of magnitude. The result is motion that is several orders of magnitude faster than c without violating the conditions and laws of either Special Relativity or General Relativity.

We are not going to 'contract space' and drag a distant star toward us, the idea is preposterous. That is, the classic demonstration of folding a piece of paper in half and making a short cut across two points by sticking a pencil through it is ludicrous. Even a supermassive Black Hole cannot warp space to that degree over such great distances. The obvious alternative is to alter our local value Lp out to extend to great distances via the applications that will be discussed in the specific engine design in this paper utilizing the Quantum Zeno Effect and Quantum Anti-Zeno Effect to alter the progression of time, thus via the demands of General Relativity alter the shape of space. As we saw with the LIGOS data, creating a space-time inversion occurs in nature, it does not require any 'exotic matter,' 'negative energy,' or any other 'exotic conditions.'

We will be extending the nose of our 'starship' out to vast distance by rendering vast changes in the values Lp and tp, then 'snapping the rear' of our 'starship' in a rather continuous fashion until we reach our destination. Since the only possible quantized motion allows for v=0 or v=c, we will in a sense be travelling at great velocity without 'moving.' For instance, we may be jumping one light year by only moving one Lp distance. Furthermore, since tp is also the finest slice of time allowable in normal space-time, we will be making that jump in 10^{-44} seconds.

We will be following the most fundamental principle of Temporal Mechanics:

Every system on every scale, from the quantum scale to cosmological scales proceeds such that:

$$t' \to \infty$$

And/or

$$t' \to 0$$

We then begin looking t the substitutions:

$$G' = 6.67384(80) \times 10^{-11} \text{ m}^3/\text{Kg (s)}^2$$

Substitute t' for s

$$G' = 6.67384(80) \times 10^{-11} \text{ m}^3/\text{Kg (t')}^2$$

$$t_p' = t_{p0} \sqrt{1 - \frac{2GM}{rc^2}}$$

$$t_p' = t_{p0} \sqrt{1 - \frac{2G'M}{rc^2}}$$

Therefore, since tp' and G' are then self-similar

$$t_p' \leftrightharpoons t_{p0} \sqrt{1 - \frac{2G'M}{rc^2}}$$

Gravitation becomes a fractal. This is described in greater detail in Temporal Mechanics 101, Bray.

Effects out to a Cosmological Scale

This sub-topic is important to understand because it explains that the evidence that the argument rendered above has been observed for over half of a century but not characterized, possibly due to misconceptions of what the observations mean. When explained, they become obvious. This all goes into the feasibility of the engine on traversing cosmological distances.

The implications for the Hubble Parameter, H_0, at very large scales, are that it introduces significant variations in G', potentially accounting for observed gravitational unexplained anomalies, due to velocity of recession.

For example, the Hubble Parameter, H_0 is currently estimated to be about 67800 m/sec*Mega-parsec (Planck Mission Data 2013), which translates to 0.020787575 (need the precision) m/s* per light year distance.

At 5 billion light years distance, the recessional velocity is about 1.039E+08, roughly 1/3 the speed of light, and t' becomes about 1.066, in which case, plugging t' back into $6.67384(80) \times 10^{-11}$ m³/Kg (t')² results in G' dropping to about 5.87×10^{-11} m³/Kg s². At 10 billion light years G' falls off to about 3.47×10^{-11} m³/Kg s². At 13 billion light years, 1.26×10^{-11} m³/Kg s², at 13.8 billion light years about 5.7×10^{-12} m³/Kg s²

Placing the age of the universe at 13781200000 years. It isn't until we get to the contrived value of 70.9 km/sMpc, which is within the margins of error for the WMAP 2007, 2009, and 2010 data, that we yield a value in agreement with observations in redshift data: (H_0=70.9 km/sMpc derived from fractal analysis)

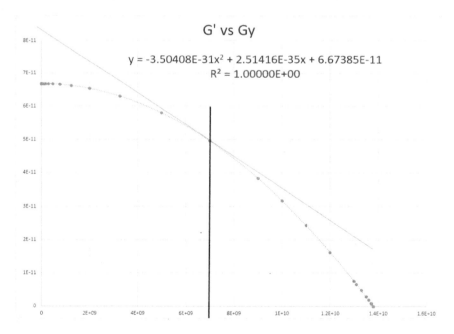

The tangent slope/intercept is described in Quantum Temporal Dynamics [Bray, Temporal Mechanics 101, 2004] as the point where the expansion of space-time makes the shift from 'coasting' to accelerating. Roughly at 7 billion years. The key element here is that G' is varying throughout the age of the cosmos, dropping to zero at the Big Bang, and leveling off just about now.

Where we defined G' as:

$$G' = 6.67384(80) \times 10^{-11} \text{ m}^3/\text{Kg } (t')^2$$

We are using recessional velocity to derive t'.

Interestingly, I nearly forgot to point out, this entire cosmos expands according to a Fibonacci sequence. Some mathematicians will argue, but I will set the record straight. As a linear progression, the Fibonacci sequence is not a fractal, as a 2-dimensional Cartesian plot the Fibonacci sequence is not a fractal. However, as a 3-dimensional *surface description* (such as this cosmos, Holographic Theory) the Fibonacci sequence ***is a fractal.*** (The Fibonacci Word is a binary issue related to computer language and not associated with this argument). There are 2-dimensional examples, however. As 3-dimensional examples, the conch shell is not just a golden spiral, but as a 3-d object, the consecutive shells are sequentially larger in radii according to the Fibonacci sequence. A sunflower not only has spirals of Fibonacci numbers, but the 3-d packing of those seeds is n/φ, the tightest possible packing possible. A conch shell, growing in 3-d in such a self-similar fashion constitutes a fractal, but not the 2-d spiral. Nonetheless, According to Falconer, Exact self-similarity: *identical at all scales; e.g. Koch snowflake*, constitutes a fractal, as the Fibonacci sequence is a recursive process:

$$F_n = F_{n-1} + F_{n-2}$$

Or

$$F_n \rightleftarrows F_{n-1} + F_{n-2}$$

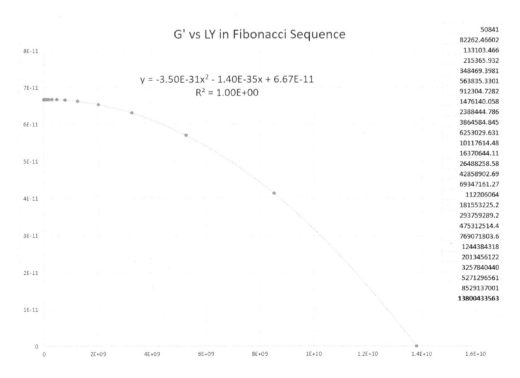

Here we have the cosmos expanding back to the Big Bang, with a Fibonacci Fractal starting at 50841 years ago (in this plot), and ending dead on accurate at 13.8 billion years with G' falling off to zero. Furthermore, the *equation of the line* is exactly the same as that generated by the sum of the data from the deep sky surveys in the graph on the previous page. The implications and details of this Fibonacci Fractal will be discussed a bit later on.

As for isolating the actual Fibonacci Fractal, the idea that a Fibonacci sequence must begin at '1' is incorrect. One merely needs to use the equation to generate the linear sequence.

This is a very interesting feature, determining G' as a function of recessional velocity, t', which in turn is a Fibonacci Fractal with a tangent intercept at the exact location where cosmologists claim the 'acceleration' due to 'Dark Energy' took over. Moreover, every value is exact without being reverse engineered.

What does this fracking of space-time expansion have to do with the engine design? It demonstrates a level of understanding of space-time physics that surpasses by far anything physics has come close to comprehending or even considering at this time. An expanding Holographic surface whose expansion characteristic is a Fibonacci Fractal of the second order equation stated above, with at least 29 fracs as shown. Actually, a frac is occurring every Planck interval. This contributes to Wheeler's foamy characteristic as space-time expands and rearranges itself, creating virtual particles and fields in the process.

We have recently (this revision is as of 13 Mar 17) discovered that the 'acceleration' issue of space-time may be an artifact of measurement. More data in the pool has reduced the certainty from 5.5 sigma to 3 sigma. Three sigma is still 99.9% certainty. However, scientists expect a trend downward. Further down I show the variation in the data collected over many years, the variation for H_0 does not vary much. However, variation in the apparent acceleration of the expansion, no matter how many sigma is explained off as: 'Variation of expansion rate due to dark energy density variation.' Nonetheless, Dark Energy has never been detected, no evidence for it exists, other than some other unexplained phenomenon, and it remains a hypothesis. Yet, it is used to explain off variation in the hard data. The failing 'accelerating expansion hypothesis,' as well as the critical density of the cosmos is 70% too lite are the best evidence to date.

Some type of artifact occurs, we call it going from coasting mode to accelerating mode in the expansion of the cosmos exactly at ½ the age of the cosmos. That is, in various papers, give or take a little, for the first ½ the age of the cosmos, the expansion during the first half is gliding apart, then at the half way point begins to accelerate. Scientists hypothesis that Dark Energy dominates at this point, even though it makes up the greater portion of mass-energy, some 68.3% (I call into question the number of significant figures here, which is exactly 1 sigma). This 0.5c is obtained by taking the value for G' (some 4.957 x 10^{-11}) circled in the data below at 7 billion years at the tangent point and plugging it back into the equations to derive velocity. This is in reasonable agreement with current estimates. Currently, the entire hypothesis is in question. What we want to look for here is an artifact.

In our quantized motion on a Planck scale this (0.5c) unambiguously translates to 1Lp/2tp, go-stop-go at v=c and v=0. If we look at this from the perspective of *information*, and as such, Pierre Nicolini's Entropic Gravitational approach to *information* given by:

$$F = \frac{GMm}{t^2}\left[1 + 4l^2p\frac{\partial s}{\partial A}\right]$$

Which Nicolini published as:

$$F = \frac{GMm}{r^2}\left[1 + 4\ell_P^2 \frac{\partial s}{\partial A}\right]$$

And I have taken the liberty to use the transformation c=l/t.

[Piero Nicolini 'Entropic force, noncommutative gravity and ungravity,' Frankfurt Institute for Advanced Studies (FIAS), August 16, 2010]

Taking this down to a quantum scale, which we can see inside of the brackets as Planck area, once the Planck areas are isolated from one another with no possible overlap, we can then go into what we see as an artifact of accelerated expansion, at exactly 1Lp/2tp, where overlap becomes impossible.

Circled is the endpoint of 13.800 billion years; the midpoint of G' = 4.95685 x 10^{-11} at 7 billion years ago, which translates to 0.5c.

For those who missed what I am doing here, I am merely converting the Hubble constant from mega-parsec to light years. Then I am converting that recessional velocity to meter per second per light year. I then plug that recessional velocity at a given distance in light years in t' according to Special Relativity:

$$t' = \frac{t_0}{\sqrt{1 - (\frac{v}{c})^2}}$$

I then take t' and plug it back into the gravitational constant:

G' = 6.67384(80)×10^{-11} m³/Kg (t')²

That is, the process of converting the Hubble constant from mega-parsec to light years, converting that recessional velocity to meter per second per light year, then plug that recessional velocity at a given distance in light years in t' according to Special Relativity:

$$t' = \frac{t_0}{\sqrt{1 - (\frac{v}{c})^2}}$$

I then take t' and plug it back into the gravitational constant: **G' = 6.67384(80)×10⁻¹¹ m³/Kg (t')²**

To arrive at a value G' which has the same end result as inverting the Hubble Parameter to Hubble Time, $1/H_0$ to arrive at an age for the cosmos share no common elements whatsoever, and there is no possibility that this is a mere mathematical artifact, but a genuine mathematical process. What the process allows us to do is to chart the progress of G' back through time to understand how the Big Bang occurred. In short, gravity did not exist, but time was dilated to near infinity.

And as it turns out, at approximately 13,800,962,991 light years G' falls off to zero. Since G', nor t', nor any of the other mathematical functions that I am using in this case are derived from one another, this is not a mathematical artifact. G' is a real value, and dependent on t', and the recessional velocity, H_0. Yet, the three values are interdependent. The data agrees with what is referred to as Hubble time, t_H, the reciprocal value of H_0, regardless of the fact that there is no relationship between the two processes.

For instance, using the Planck Mission data at H_0 = 67.8m/sMpc we have:

$t_H = 1/H_0 = 1/67.8$km/sMpc $= 4.55 \times 10^{17}$ sec $= 14.4$ billion years

1. Bucher, P. A. R.; et al. (Planck Collaboration) (2013). "Planck 2013 results. I. Overview of products and scientific Results". arXiv:1303.5062 [astro-ph.CO].
2. "Planck reveals an almost perfect universe". ESA. 21 March 2013. Retrieved 2013-03-21.
3. "Planck Mission Brings Universe Into Sharp Focus". JPL. 21 March 2013. Retrieved 2013-03-21.
4. Overbye, D. (21 March 2013). "An infant universe, born before we knew". New York Times. Retrieved 2013-03-21.
5. Boyle, A. (21 March 2013). "Planck probe's cosmic 'baby picture' revises universe's vital statistics". NBC News. Retrieved 2013-03-21

Which is somewhat higher than the observed 13.799(+/-)0.021 billion years. Using the H_0 derived from the process I have just described, we have arrived at:

$t_H = 1/H_0 = 1/70.9$km/sMpc $= 4.35 \times 10^{17}$ sec $= 13.8$ billion years

1. Planck Collaboration (2015). "Planck 2015 results. XIII. Cosmological parameters (See Table 4 on page 31 of PDF).". arXiv:1502.01589.
2. C. R. Lawrence, JPL, for the Planck Collaboration, Astrophysics Subcommittee, NASA HQ (18 March 2015) "Planck 2015 Results" (See page 29 of pdf) http://science.nasa.gov/media/medialibrary/2015/04/08/CRL_APS_2015-03-18_compressed2.pdf

3. http://www.teachastronomy.com/astropedia/article/Precision-Cosmology

This process has also nailed the Hubble constant down to an exact figure independent of direct measurement, whereas the direct measurement approach had H_0 in a range from the most recent Planck mission (which, incidentally disagreed with all of the historical data) at 67.8km/sMpc to the Chandra Mission at 92km/sMpc. Depending on how you approach your statistics, that is greater than one standard deviation, meaning that the history of the data is unrelated from one mission to another, AKA, nearly random data. Therefore, when we approach the subject of quantized redshift data on a cosmological scale, the rebuff coming from astrophysicists, who are in turn using this random Hubble and other data as a scaffold for their arguments, is dismissed. If we look at the data sets collected over the years, the correlation of such a sensitive test as redshift is not very impressive:

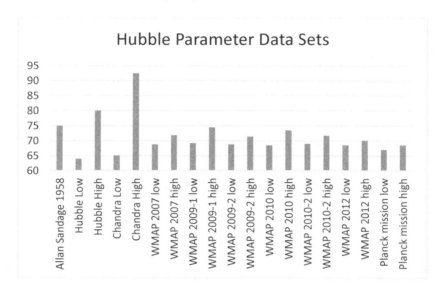

In fact, if we take a look at the range of the raw data as presented in the peer reviewed data, we only have 5% correlation: (A Nobel was handed out for this 'accelerating expansion based on I think it was the Chandra data, which is way out)

	x	x-mean	(x-mean)^2
Allan Sandage 1958	75	3.52	12.3904
Hubble Low	64	-7.48	55.9504
Hubble High	80	8.52	72.5904
Chandra Low	65.1	-6.38	40.7044
Chandra High	92.5	21.02	441.8404
WMAP 2007 low	68.8	-2.68	7.1824
WMAP 2007 high	71.9	0.42	0.1764
WMAP 2009-1 low	69.2	-2.28	5.1984
WMAP 2009-1 high	74.5	3.02	9.1204
WMAP 2009-2 low	68.8	-2.68	7.1824
WMAP 2009-2 high	71.4	-0.08	0.0064
WMAP 2010 low	68.5	-2.98	8.8804
WMAP 2010 high	73.5	2.02	4.0804
WMAP 2010-2 low	69	-2.48	6.1504
WMAP 2010-2 high	71.7	0.22	0.0484
WMAP 2012 low	68.52	-2.96	8.7616
WMAP 2012 high	70.12	-1.36	1.8496
Planck mission low	67.03	-4.45	19.8025
Planck mission high	68.57	-2.91	8.4681
		sum	710.3838
		sum/18	39.46576667
		stdev	6.282178497
		rsd	15.918045

Forgive all the decimals, I just tend to leave my Excel set that way. A standard deviation of 6 in a range of 28 is a bit on the high side, almost 25% of our range, with a Relative Standard Deviation of 16%. That is a very high RSD. To put it in perspective, that much variation in a motor part production would shut the production line down. That much variation in a drug's bioavailability would have it pulled from market. That much variation in the stock market would be economic near chaos. That much variation in the value of the dollar would be total economic collapse.

1. Riess, A.; et al. (September 1998). "Observational Evidence from Supernovae for an Accelerating Universe and a Cosmological Constant". The Astronomical Journal 116 (3): 1009–1038. arXiv:astro-ph/9805201. Bibcode:1998AJ....116.1009R. doi:10.1086/300499.
2. Perlmutter, S.; et al. (June 1999). "Measurements of Omega and Lambda from 42 High-Redshift Supernovae". The Astrophysical Journal 517 (2): 565–586. arXiv:astro-ph/9812133. Bibcode:1999ApJ...517..565P. doi:10.1086/307221.
3. Coles, P., ed. (2001). Routledge Critical Dictionary of the New Cosmology. Routledge. p. 202. ISBN 0-203-16457-1.
4. "Hubble Flow". The Swinburne Astronomy Online Encyclopedia of Astronomy. Swinburne University of Technology. Retrieved 2013-05-14.
5. Lemaître, G. (1927). "Un univers homogène de masse constante et de rayon croissant rendant compte de la vitesse radiale des nébuleuses extra-galactiques". Annales de la Société Scientifique de Bruxelles A (47): 49–59. Bibcode:1927ASSB...47...49L. Partially translated in Lemaître, G. (1931). "Expansion of the universe, A homogeneous universe of constant mass and increasing radius accounting for the radial velocity of extra-galactic nebulae". Monthly Notices of the Royal Astronomical Society 91: 483–490. Bibcode:1931MNRAS..91..483L. doi:10.1093/mnras/91.5.483.
6. van den Bergh, S. (2011). "The Curious Case of Lemaitre's Equation No. 24". Journal of the Royal Astronomical Society of Canada 105 (4): 151. arXiv:1106.1195. Bibcode:2011JRASC.105..151V.

7. Block, D. L. (2012). "Georges Lemaitre and Stiglers Law of Eponymy". In Holder, R. D.; Mitton, S. Georges Lemaître: Life, Science and Legacy. Astrophysics and Space Science Library 395. pp. 89–96. arXiv:1106.3928. Bibcode:2012ASSL..395...89B. doi:10.1007/978-3-642-32254-9_8. ISBN 978-3-642-32253-2.
8. Reich, E. S. (27 June 2011). "Edwin Hubble in translation trouble". Nature News. doi:10.1038/news.2011.385.
9. Livio, M. (2011). "Lost in translation: Mystery of the missing text solved". Nature 479 (7372): 171. Bibcode:2011Natur.479..171L. doi:10.1038/479171a.
10. Livio, M.; Riess, A. (2013). "Measuring the Hubble constant". Physics Today 66 (10): 41. Bibcode:2013PhT....66j..41L. doi:10.1063/PT.3.2148.
11. Hubble, E. (1929). "A relation between distance and radial velocity among extra-galactic nebulae". Proceedings of the National Academy of Sciences 15 (3): 168–73. Bibcode:1929PNAS...15..168H. doi:10.1073/pnas.15.3.168. PMC 522427. PMID 16577160.
12. Longair, M. S. (2006). The Cosmic Century. Cambridge University Press. p. 109. ISBN 0-521-47436-1.
13. Bucher, P. A. R.; et al. (Planck Collaboration) (2013). "Planck 2013 results. I. Overview of products and scientific Results". arXiv:1303.5062 [astro-ph.CO].
14. "Planck reveals an almost perfect universe". ESA. 21 March 2013. Retrieved 2013-03-21.
15. "Planck Mission Brings Universe Into Sharp Focus". JPL. 21 March 2013. Retrieved 2013-03-21.
16. Overbye, D. (21 March 2013). "An infant universe, born before we knew". New York Times. Retrieved 2013-03-21.
17. Boyle, A. (21 March 2013). "Planck probe's cosmic 'baby picture' revises universe's vital statistics". NBC News. Retrieved 2013-03-21.
18. Bennett, C. L.; et al. (2013). "Nine-year Wilkinson Microwave Anisotropy Probe (WMAP) observations: Final maps and results". The Astrophysical Journal Supplement Series 208 (2): 20. arXiv:1212.5225. Bibcode:2013ApJS..208...20B. doi:10.1088/0067-0049/208/2/20.
19. Jarosik, N.; et al. (2011). "Seven-year Wilkinson Microwave Anisotropy Probe (WMAP) observations: Sky maps, systematic errors, and basic results". The Astrophysical Journal Supplement Series 192 (2): 14. arXiv:1001.4744. Bibcode:2011ApJS..192...14J. doi:10.1088/0067-0049/192/2/14.
20. Results for H0 and other cosmological parameters obtained by fitting a variety of models to several combinations of WMAP and other data are available at the NASA's LAMBDA website.
21. Hinshaw, G.; et al. (WMAP Collaboration) (2009). "Five-year Wilkinson Microwave Anisotropy Probe observations: Data processing, sky maps, and basic results". The Astrophysical Journal Supplement 180 (2): 225–245. arXiv:0803.0732. Bibcode:2009ApJS..180..225H. doi:10.1088/0067-0049/180/2/225.
22. Spergel, D. N.; et al. (WMAP Collaboration) (2007). "Three-year Wilkinson Microwave Anisotropy Probe (WMAP) Observations: Implications for cosmology". The Astrophysical Journal Supplement Series 170 (2): 377–408. arXiv:astro-ph/0603449. Bibcode:2007ApJS..170..377S. doi:10.1086/513700.
23. Bonamente, M.; Joy, M. K.; Laroque, S. J.; Carlstrom, J. E.; Reese, E. D.; Dawson, K. S. (2006). "Determination of the cosmic distance scale from Sunyaev–Zel'dovich effect and Chandra X-ray measurements of high-redshift galaxy clusters". The Astrophysical Journal 647: 25. arXiv:astro-ph/0512349. Bibcode:2006ApJ...647...25B. doi:10.1086/505291.
24. Freedman, W. L.; et al. (2001). "Final results from the Hubble Space Telescope Key Project to measure the Hubble constant". The Astrophysical Journal 553 (1): 47–72. arXiv:astro-ph/0012376. Bibcode:2001ApJ...553...47F. doi:10.1086/320638.
25. Overbye, D. (1999). "Prologue". Lonely Hearts of the Cosmos (2nd ed.). HarperCollins. p. 1ff. ISBN 978-0-316-64896-7.

26. Sandage, A. R. (1958). "Current problems in the extragalactic distance scale". The Astrophysical Journal 127 (3): 513–526. Bibcode:1958ApJ...127..513S. doi:10.1086/146483

If any of these papers were kind enough to present the region of sky they mapped during the measurements in question with any consistency we could answer a huge spectrum of questions. In any case, we can merely look at the difference between measurements without actually knowing the distances involved and see a distinct pattern emerge of quantization:

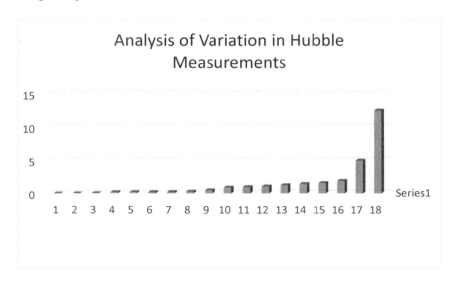

Interestingly, the very tall spikes belong to Chandra and Hubble, the only two of the measurement tools capable of measuring out to the very edge of the Visible Cosmos. We see a grouping 1-3, then 4-6, 7-9, then another grouping 10-15, then 16-18 show deep space measurements, redshift near black. Keep in mind that many scientists published papers noting quantizations in the data and the Orthodoxy scoffed at them.

For group 1-3 the correlation is 98.7%

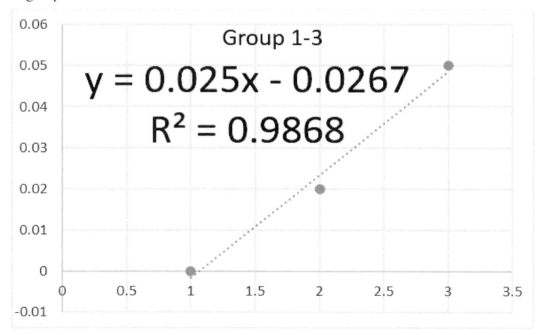

For group 4-6 the correlation is 100%

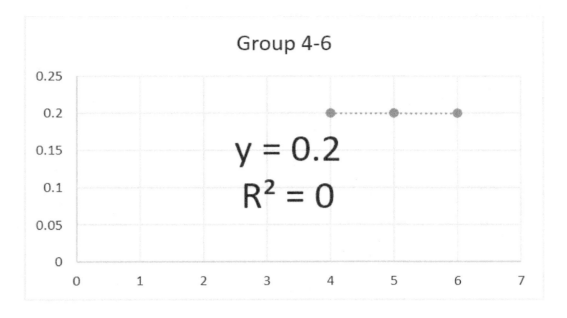

Group 7-9 correlation is 93%

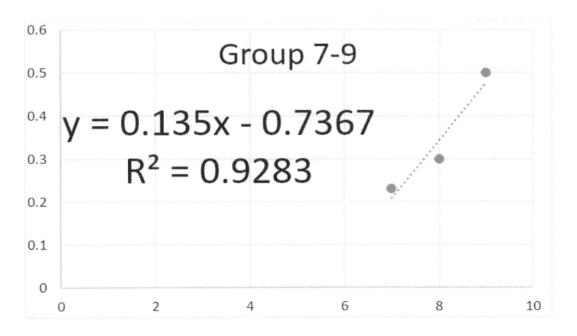

Group 10-15 is 98% correlation:

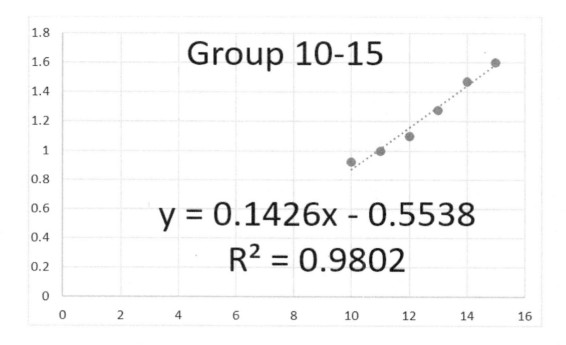

And even group 16-18 is 95% correlated:

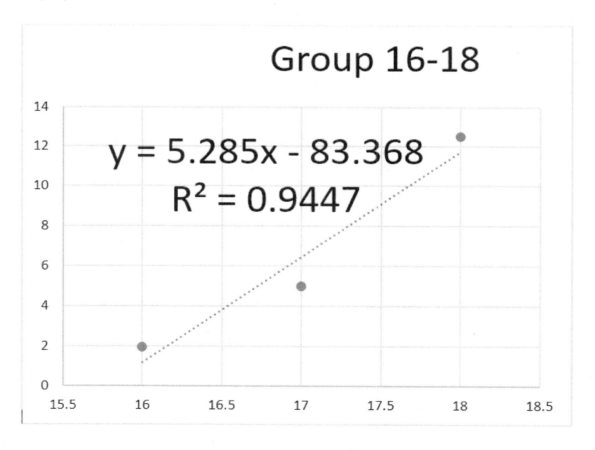

The problem isn't measurement. The problem is the Fibonacci Fracking on a cosmological scale. If a mission survey is looking at a particular depth of sky, they will in turn be looking into a 'frac' of space-time and report a different recessional value.

Just as a Mandelbrot set has macroscopic features and microscopic features depending on how you 'zoom' in this is also true of space-time.

The actual quantization shows up as the slope in each grouping, and each grouping has a rather tight correlation. In fact, given that the correlation of the Hubble Parameter data is in itself not well correlated using the most sophisticated technology available, it is safe to say that this high degree of correlation I am displaying here describes why the Hubble Parameter data is so uncorrelated. The uncorrelation in the Hubble Parameter search is dependent on which portion and the depth of sky viewed in order to achieve such data. The proof of this is that I just reverse engineered said data to an average of 95% correlation in each case by observing the quantization (even though I lack distance parameters, I reverse engineered around the need for it) of the data groupings. That is, the correlation of the various Hubble Parameter searches over the past half century is perhaps 75% (by eye), whereas my quantization data for redshift (Hubble Parameter quantized to cosmological scales outward from the Big Bang) averages 95%. This means that any scientist bringing a 'smooth' redshift argument to the table has 6% certainty to offer, whereas I have 95% certainty to offer in my quantized redshift data, using their own raw data, *correctly*.

The idea that the subatomic does not scale up to the macroscopic has no support by any evidence or experiment, ever, under any conditions in any lab in human history. In fact, the discovery by Benoit Mandelbrot of Fractal Mathematics has taught us that nature scales everything from the infinitesimal to the universal as a means of *conserving information* by means of the fractal. In nature, this is always the case, without exception.

And that is called the -1 law of thermodynamics: the conservation of information. *And we can use this property, in part given to us by Leonard Susskind and in part by Benoit Mandelbrot, to use nature to our advantage, whatsoever we may apply on a microscopic, Planck scale, we can scale up successfully to the macroscopic, because nature conserves information by utilizing the same information paradigms on every scale.*

It is important to note that this subtopic is important to understand because in order to making a working engine we must scale the QZE up from the subatomic up to the macroscopic scale, and we must have a valid hypothesis in place for doing so before we proceed.

Another side note is that the fate of the universe, according to this relationship which indicates a G' that is increasing toward infinity (asymptotically) as time (the age of the cosmos) increases would seem to indicate that recollapse is not an inevitability. G' seems to be levelling off at this time (if you look at the graph a few pages back), yet the cosmos continues to expand. Heat Death seems to be the fate of the universe.

With respect to the apparent accelerated expansion of the universe, this premise is based upon Friedmann's equations:

$$H^2 \equiv \left(\frac{\dot{a}}{a}\right)^2 = \frac{8\pi G}{3}\rho - \frac{kc^2}{a^2} + \frac{\Lambda c^2}{3},$$

Where H is the Hubble parameter, a is the scale factor, G is the gravitational constant, k is the normalized spatial curvature of the Universe and equal to −1, 0, or +1, ρ is the mass-energy density of the universe, 'a' is the scale factor, and Lambda is the cosmological constant.

If G' increases, H increases, meaning that the recessional velocity increases. This seems counter intuitive. One would think that if the Gravitational Constant increases space-time would be held together or bound more tightly by gravitation. However, the G' factor and H in this equation show the recessional velocity increasing with time. The counter intuitive rationale for this is that the reason G' increases with time is that G' is a function of t':

$$G' = 6.67384(80) \times 10^{-11} \text{ m}^3/\text{Kg } (t')^2$$

And in turn, t' is a function of gravimetric time dilation, but t' is falling as the universe expands, leaving G' to increase:

$$t' = t_0 / \sqrt{1 - \frac{2GM}{rc^2}}$$

Furthermore, there is a self-similar (fractal) occurring:

$$\text{Since } G' = 6.67384(80) \times 10^{-11} \text{ m}^3/\text{Kg } (t')^2$$

And

$$t' = t_0 / \sqrt{1 - \frac{2G'M}{rc^2}}$$

Then

$$t' \rightleftharpoons t_0 / \sqrt{1 - \frac{2G'M}{rc^2}}$$

For instance, the Swarzschild Radius of the mass-energy (observable) of the universe is approximately 13.7 billion light years.

1. Valev, Dimitar (October 2008). "Consequences from conservation of the total density of the universe during the expansion". arXiv:1008.0933 [physics.gen-ph].

2. Deza, Michel Marie; Deza, Elena (Oct 28, 2012). Encyclopedia of Distances (2nd ed.). Heidelberg: Springer Science & Business Media. p. 452. doi:10.1007/978-3-642-30958-8. ISBN 978-3-642-30958-8. Retrieved 8 December 2014.

At that time, t' was approaching infinity. Since that time, space-time has been expanding, because G' was zero, t' has been falling off to some tangible value, and G' has been steadily increasing as shown in the graphs rendered above. As G' has been steadily increasing as a result of t' decreasing, H_0 has been increasing. I will therefore render H the variable H_0'. And we will say that H_0' is directly related to G' by:

$$H^2 \equiv \left(\frac{\dot{a}}{a}\right)^2 = \frac{8\pi G}{3}\rho - \frac{kc^2}{a^2} + \frac{\Lambda c^2}{3},$$

In which case we can say:

$$H_0' = \sqrt{\frac{8\pi G'}{3}\rho - \frac{kc^2}{a^2} + \frac{\Lambda c^2}{3}}$$

The lambda term is not vital to the equation. In any case, H_0' increases with time as G' increases because t' decreases as gravimetric time dilation decreases as the cosmos expands away from the Swarzschild Radius (13.7 billion light-years).

That is, what the artifact is, the apparent acceleration of the expansion of the cosmos, is a decrease in gravimetric time dilation.

Nothing in particular happens with the velocity of recession at midpoint:

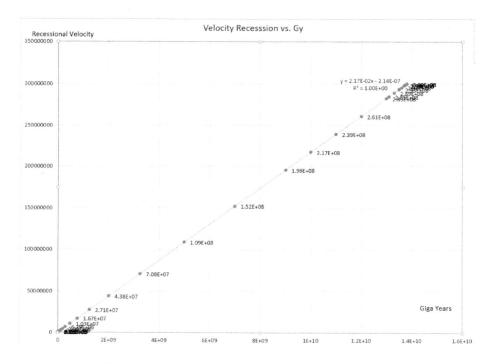

At this point t' = 1.16034, G' = 4.95685E-11. However, since the fractal is of the sort:

$$t' \rightleftharpoons t_0 / \sqrt{1 - \frac{2G'M}{rc^2}}$$

And thus:

$$H'_0 \rightleftharpoons \sqrt{\frac{8\pi G'}{3} p - \frac{kc^2}{a^2} + \frac{\Lambda c^2}{3}}$$

This is the source of the curve with the intercept right where physicists hypothesize two things they can neither see nor substantiate with any evidence whatsoever, Dark Matter and Dark Energy:

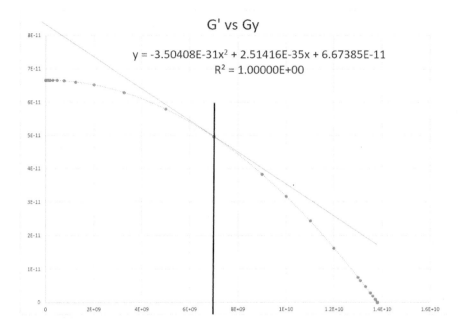

Nor can they produce a curve or series of curves such as I have provided in this paper thus far, *to be perfectly clear on that issue.* They instead introduce words where none existed before…

H_0 is a fractal of G', is a fractal of t'. Just to be clear, although it appears the recessional velocity passes through zero *now,* that is an artifact of distance. That is, at *zero distance*, the recessional velocity is zero.

As we say with respect to *information entropy* in Nicolini's approach to Gravitation:

$$F = \frac{GMm}{t^2}\left[1 + 4l^2 p \frac{\partial s}{\partial A}\right]$$

Note first that Nicolini does not refer to entropy as a final state as does Holography.

This approach ignores the current apparent 'flatness' of space-time and considers it (flatness) only as a state of entropy. We can rearrange Nicolini's equation:

$$t' \leftrightarrows \sqrt{\frac{G'mm}{F}\left[1 + 4l^2 p \frac{\partial s}{\partial A}\right]}$$

Thus, we see an increasing recessional velocity with time, *an acceleration in the expansion of space-time.* There is no 'Dark Energy' requirement here, only the natural relationships between the irrefutable presence of gravimetric time dilation that *must exist* in an expanding cosmos, its effect on G', and its consequent effect on the measurement of H_0'. Given that G is time dependent, there is no option for G to not be affected by gravimetric time dilation, and hence, affect in turn the Hubble Parameter.

The argument that t' is insignificant is insubstantial because typically t' is only applied to a local environment whereas H_0 is applied to large scale space. H_0 even on a scale of a single light-year translates to 0.020787575 m/s* per light year distance. On 1 meter scale this translates to 2.2×10^{-18} m/s/m. If one considers the amount of Gravitation necessary to hold a galactic cluster together, t' becomes significant. This type of calculation requires supercomputers I currently have no access to, or I would render the data. Basically something like 6×10^{11} solar masses, about 2000 light-years thick and roughly 150,000 light-years across. The value t' therefore varies considerably throughout.

Notably, however, giving rise to the 'Dark Matter' myth is the fact that rather than spin like a soup of stars around the center, it spins more like a record album, with the outermost stars only lagging a bit, but much less than they should as a gas or liquid. The problem is with invisible mass is that such mass would reshape the galaxy, particularly at the thinnest regions of the disk, as this 'Dark Matter' is proposed to exist as a spherical halo around the Milky Way. If a sphere of mass is equally portioned around the disk of the Milky Way, it would have long ago torn the disk feature into a nebulous galaxy. The Dark Matter hypothesis is grade school wrong.

However, if we replace the absurdity of Dark Matter with entropic gravity:

$$F = \frac{GMm}{t^2}\left[1 + 4l^2 p \frac{\partial s}{\partial A}\right]$$

We find that this equation is a surface phenomenon. That is, is doesn't see a planet as a massive sphere, the gravity is the surface of space-time facing you. That is, the moon is not affected by the side of the Earth it does not see, it is affected by the side of the Earth facing it. Gravity then, is a surface phenomenon, which essentially agrees with Holographic Theory.

This is not a suggestion that gravitational lensing is a slight of hand, merely that the same principle that Holographic Theory claims our 4-dimensional façade is a 2-dimensional Schwarzschild surface applies to gravity as an entropic phenomenon rather than a Dark Matter solution which fails on a grade school level as it would have torn the galaxies structure to a nebula by now.

All of this discussion regarding cosmological scales and forms of gravitation, the nonsense of Dark Matter and Dark Energy is for the edification of the reader in understanding the engine design on *what I consider a mechanical level.* What traveling without moving across fractalized space-time, a surface where time is infinitely dilated existing as a standing Planck Interval (not a flow) where *information* is preserved by photons using a manifold that is superpositioning Lp to any distant point, technically arriving in the future, means. We'll discuss the actual mechanism later. It is as simple as a wind-up toy once explained.

Another way to think of the phenomenon is that t' is anti-dilating, that is, speeding up (like our space-time inversion). This is much the same as sitting from the preferential perspective of inside Earth's gravity well observing the space shuttle and noting that their clocks are running slightly faster than here on the ground. In this case, we are observing from some preferential perspective noting the tick rate of the recessional velocity, which is not in our preferential perspective. It may be that the very scaffold of our cosmos is contained within a larger system altogether, and this notion is aligned with the Inflation Model to some degree. In any case, what we are looking at is the tick rate of t' quickening, thus an observation of apparent acceleration in the rate of expansion of space-time. For all practical purposes, it is not possible for t' on a cosmological scale to have the same value now as it did 13.7 billion years ago, and it is not possible that it will have the same value a billion years from now that it has today. The value t' on a cosmological scale must be changing, that point is irrefutable.

Looking at the equation:

$$H_0' \leftrightharpoons \sqrt{\frac{8\pi G'}{3}\rho - \frac{kc^2}{a^2} + \frac{\Lambda c^2}{3}}$$

In order for H_0' to be quantized, G' must be quantizing to some astronomical value. In order for G' to be quantizing to some astronomical value, t' must be quantizing to some astronomical value:

G' = 6.67384(80)×10⁻¹¹ m³/Kg (t')²

The value t' is being derived from gravimetric time dilation (which is falling as the universe expands, leaving G' to increase without bound, albeit, asymptotically):

$$t' = t_0 / \sqrt{1 - \frac{2GM}{rc^2}}$$

Given that:

$$tp' = \sqrt{\frac{hG'}{2\pi c^3}}$$

Our time dilation factor can be expressed as a fractal:

Since

G' = 6.67384(80)×10⁻¹¹ m³/Kg (t')²

Then

$$t' \lessgtr t'_p / \sqrt{1 - \frac{2G'M}{rc^2}}$$

I do not have the computer power or software to resolve the fractal here. However, this fractal equation states the astronomical quantization of H_0' via the fractalization of G'. The quantization is unique in that as a fractal it is not a set size, but changes with each fractation, which is why it has been so elusive. You will also note that in the graphs of the quantizations I have shown above, the quantizations are described by their slopes, which are not equal to one another. This value G' at each frac is fed back into our equation:

$$H'_0 \rightleftarrows \sqrt{\frac{8\pi G'}{3}p - \frac{kc^2}{a^2} + \frac{\Lambda c^2}{3}}$$

Given that H_0 is in units of m/s* ly, we can say H_0 is in units of m/t' *ly as the value t' changes as the cosmos expands:

This self-similarity and interdependence between $t' \rightleftarrows G' \rightleftarrows H_0$ leads us to:

$$H'_0 \rightleftarrows \sqrt{\frac{8\pi G'}{3}p - \frac{kc^2}{a^2} + \frac{\Lambda c^2}{3}}$$

And a new value for H_0' is obtained, different from the previous value as time progresses and space-time flattens. Considering space-time as 'flat,' we consider the clustering and superclustering of galaxies regions where $t' \neq 1$. Gravitational lensing is another example of the variable non-flatness of space-time on large localized scales.

It is typical with fractals that each iteration is smaller in size than the previous, but there is no rule on a cosmological scale that the pattern may not be atypical (larger with each iteration), given that the universe started as a zero dimensional point, in theory. Thus, the quantizations observed by many astrophysicists are indeed large scale fractals as indicated by the mathematical proofs and the simple arrangement of the data graphs above.

The simple bar graph I showed organizing the quantizations into groups displays this characteristic a bit more clearly:

Analysis of Variation in Hubble Parameter Measurements

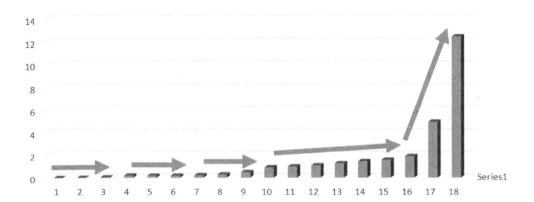

Showing 4 fracs in H_0. This is the tentative explanation for the high degree of correlation of the data when presented as such. Sudden shifts; usually referred to as quantizations of the Hubble constant have been noted for decades and invariably rebuffed by the Orthodoxy who do not organize the data and thus carelessly dismiss data. However, a sudden shift from random data in a Hubble parameter that is in itself a conglomerate of random data simply agreed upon by what is most recent (or otherwise have no agreed upon Hubble value), vs. 95% correlation of *all* the collected data over half of a century by mere organization by fractalization coefficients speaks volumes.

The last three may be three individual fracs given their extreme slope, for a total of perhaps seven fracs. Then, of course, we may be looking at 18 individual fracs. The fact is, in fractals, each iteration is a frac. An iteration of fracs may or may not group together to form a recognizable pattern on that scale, such as with the common Mandelbrot set where we have the realm of the beatle, the realm of the sea horses, and so on (shown below).

If they are 18 individual fracs then they are merely grouped by their slope characteristics. The slopes yield information regarding shifts in the frac characteristics, that is, large jumps in the fractal. We see such things in the sudden shift in appearance in fractals as simple as the original Mandelbrot set. Shown below is a portion of a Mandelbrot set to demonstrate the characteristics of the sudden and unexpected change that can occur on a single iteration of scale of such a fractal. The point is that if the H_0' follows a fractal pattern, then it is likely that a pattern is elusive.

For instance, one can gaze at the Mandelbrot set shown below indefinitely, and indeed engage the most powerful computer algorithms on the most powerful computers in the world available at the NSA and never, ever reverse engineer that visible pattern back to its original simple equation:

$$Z \rightleftharpoons Z^2 + C$$

All the mayhem shown below is the result of that little equation. Thus far, to the best of my knowledge, the equation has been taken out to 10^{270} iterations, and it never repeats. The 'realm of the black beetle' in the center shows up in various forms, but is never exactly the same twice.

These are observed fracs, exactly how many there are is unknown and can only be determined by which portion and most importantly, depth of sky that is observed. With a test as sensitive as redshift, which can measure the growth of a blade of grass in real time, it seems unlikely to have such a range spanning 64 to 92 km/sec*Mega-parsec (a change of nearly 50% absolute) using technology developed in just the past ten years alone, unless there is genuinely a difference between the recessional velocities of deep sky objects. It is important to keep in mind that none of these missions did a sweep of the entire sky. Each mission focused on a unique patch and depth of sky, and got a unique result.

Of course, we may be looking at a convoluted fractal of the form:

Eq 1

$$tp' \rightleftharpoons \sqrt{\frac{hG'}{2\pi c^5}}$$

Eq 2

$$t' \rightleftharpoons t_0 / \sqrt{1 - \frac{2G'M}{rc^2}}$$

Eq 3

$G' = 6.67384(80) \times 10^{-11}$ m³/Kg·(t')²

As G' to climbs toward infinity in Eq 1, tp' climbs toward infinity. As G' falls to zero, tp' falls to zero. Keep in mind that tp represents t_0, functionally.

However, in Eq 3, G' falls toward zero if tp' climbs toward infinity.

The denominator in Eq2 is interesting in that the best we can achieve in the denominator is zero. Since $2GM/c^2$ represents the Schwarzschild radius, where time infinitely dilates:

$$r_s = \frac{2GM}{c^2}$$

Keeping note that I invert the gravitational time dilation equation so as to represent *dilation* with respect to an observer outside of the gravity well consistent with dilation as defined in Special Relativity. In this case Eq2 becomes:

$$t' \leftrightarrows \frac{t_0}{\sqrt{1 - \frac{r_s}{r}}}$$

This interesting feature allows us to be a distance from a forming Black Hole, unless $r_s/r=1$, at which point, t' becomes infinite. Obviously, r_s/r cannot be less than 1, or that would place us inside of the forming Black Hole, which has no interior.

In Eq 2 as G' climbs t' climbs toward infinity, but t_0 (a variable of tp) is also climbing toward infinity. However, as G' falls to zero t' falls to t_0. It is this interdependent play that creates such seeming paradoxes. However, Eq3 states that as t' increases G' must decrease, which is in direct opposition to Eq2.

Although this is noteworthy, it is the cause of the second order curve of G' vs. Ly distance and the point of the intersection of the tangent slope. The effect G' has on t' is not a direct linear relationship in Eq2 as is Eq3.

For instance, Eq 2 has t' increasing as the cosmos expands (gravimetric time dilation), again leaving G' to increase without bound, but starting from zero. This particular feature is the artifact I spoke of which leaves us with the apparent vision of accelerating expansion. In relativistics, according to my local watch, which is dilated with respect to the early universe, I perceive the early universe as 'sped up.' Thus, the deeper I look into space by relativistic means (red-shift) the faster it seems I am receding from the source. However, if I look by standard candles, this does not appear to be the case because the supernova approach is not affected by relativistics.

In this case, G' cannot fall without t' also falling, as well as tp (t_0). It is a one way street.

Note that t_0 is t_p functionally. Therefore the total fractal course is:

$$H_0 \leftrightarrows G' \leftrightarrows tp \leftrightarrows t' \leftrightarrows r$$

Where the value *r* can represent any distance Lp on this *hypothetical* 2-dimensional Schwarzschild surface.

The value tp' is on a Planck scale, and at one time encompassed the age of the entire cosmos. The value t' is the cosmological clock, which is slowly dilating as the cosmos expands, making H_0' appear to accelerate as G' increases with time. Hence, again, we have an 'accelerating expansion' of space-time, according to Fredimann's equations.

The discrepancy appears as:

$$G' = 6.67384(80) \times 10^{-11} \text{ m}^3/\text{Kg } (t')^2$$

If time is dilating as the universe expands, then G' should be decreasing. What we don't see is t' expanding to infinity, such as a Black Hole, re-collapse of the cosmos, but leveling off. The limiting factor is that the t' fractal has reached the Planck limit:

$$tp' \rightleftharpoons \sqrt{\frac{hG'}{2\pi c^5}}$$

Given that t_0 is tp'

$$t' \rightleftharpoons t_0 / \sqrt{1 - \frac{2G'M}{rc^2}}$$

That isn't suggesting that the Planck value has been changing for the past 14 billion years, only that the fractal has been approaching the Planck limit.

Here we have quantized, Planck time, tp' and common time, t', both following the same course. The next obvious question is, 'does common time, t', alter because tp' changes?' Rather than assign a causal relationship it is safer to say that it is sufficient that common time and quantized time are doing the same thing, changing to the same degree in the same direction as a result of the same variable, G'. And, interestingly, G' is a recursive value for each t' and tp' shown above in their fractal form.

Therefore, to reiterate and summarize, what causes the quantization of recessional velocity, and furthermore, what would cause such recessional velocity both increase and to frac?

$$tp' \rightleftharpoons \sqrt{\frac{hG'}{2\pi c^5}}$$

$$t' \rightleftharpoons t_0 / \sqrt{1 - \frac{2G'M}{rc^2}}$$

$$H'_0 \rightleftharpoons \sqrt{\frac{8\pi G'}{3}p - \frac{kc^2}{a^2} + \frac{\Lambda c^2}{3}}$$

Time is quantized slave to G' (eq 1). The value t' decreases as the cosmos expands (gravimetric time anti-dilation eq 2) causing G' to increase and therefore H_0' to increase (eq 3) Evidenced by:

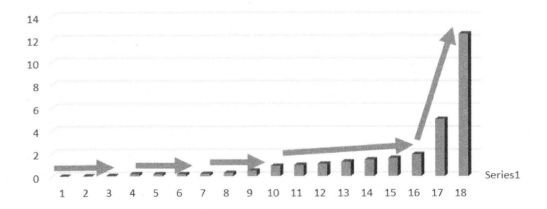

A total of 18 fracs grouped into 4 groupings detected to date as reported by the last 18 missions (sky surveys). This is not new news. Scientists have noted what they think to be large scale quantizations in the Hubble parameter for decades, but the Orthodoxy dismisses such claims with no evidence or sequitur discussion. Here it is, displayed. I will show the exact quantization a bit further on.

Furthermore, the rate of inflation accelerating that has been observed is direct evidence that this universe is contained within a larger system, regardless of the t' factor increasing.

I use the bike on a train explanation. A bike (representing our cosmos) is riding on top of a train (the larger system), and both the bike and train are decelerating, the train decelerating harder because of its immense size and age. If one were looking at the top of the train from the bike, it would appear as though we (the bike, our cosmos) are accelerating. This is the result of the bike not decelerating as hard as the train. It is only by looking at the ground that one can determine that both systems are actually decelerating. Nonetheless, we have not determined what the ground state for both systems is.

We have an inflation model, which may or may not explain the larger system in question. If the inflation model explains the larger system in question, then we can go on to say that, the inflation occurred at many orders of magnitude greater than light speed, in which case its deceleration would certainly be far greater than any possible deceleration of the subluminal deceleration of the visible cosmos. Hence, our bike on a train model has strong validity.

If the 'train' is indeed another universe, a larger universe in which our visible cosmos along with our inflationary cosmos is contained, then we have the smoking gun of a true Multiverse system. This particular model would most well conform to Tegmark's classification of a Level II Multiverse. We also have a difficult issue in defining what we can label as the ground state for both systems. The larger universe in which our cosmos is contained would be 'the top of the train.' The 'train' is braking harder because of its immense size and age. Our visible cosmos appears to be accelerating because we are looking down at the 'top of the train.' Even though our system's expansion is also decelerating, it is decelerating at a lesser rate compared to the 'top of the train.'

It may be possible to reverse-engineer the approximate size and mass-energy of the larger system based on estimated values for the mass-energy of our local cosmos and the resulting expected rate of deceleration vs. the observed, apparent 'acceleration' of the expansion.

$$\frac{RL_0}{RC_0} \alpha \frac{SL_0}{SC_0}$$

Here, RL_0 is the rate of expansion of the larger system, RC_0 is the rate of expansion of our local cosmos, SL_0 is the size, mass-energy of the larger system, and SC_0 is the size, mass-energy of our local cosmos. Which is an over generalized statement that the rate of expansion of the local observable universe as compared to the rate of expansion of the larger system in which it is contained is proportional to the size and mass-energy of each system, or rather, the ratio of the proposed sizes and mass energy they contain. By looking at the apparent acceleration of our expansion, knowing our approximate mass-energy and size and rate at which the expansion should be decelerating, we can reverse engineer the size and mass-energy of the larger system.

This equation rearranges to:

$$\frac{RL_0}{SL_0} = \frac{RC_0}{SC_0}$$

We only know the values RC_0/SC_0. However, by determining the expected deceleration of expansion of our local cosmos vs. the noted acceleration yields RL_0 of the larger system in the form:

$$\mathbf{R_{observed} - R_{expected} = RL_0}$$

In addition, since we know the ratio and values of RC_0/SC_0 we can then glean out the value SL_0. To actually determine the size and mass-energy of the larger system in which we are contained is history in the making. Not to mention *proving* one leg of the Multiple Universe Theory (Tegmark classification Level II Multiverse) and what form it takes.

Admittedly, this seems to indirectly suggest that the vast 95% of the mass-energy of the cosmos coming from Dark Energy and Dark Matter may in fact be the result of:

$$\frac{RL_0}{RC_0} \alpha \frac{SL_0}{SC_0}$$

I am not suggesting that the Multiverse is a mere 20 times larger than our cosmos, only that our cosmos is in direct contact with that much Multiverse. Like dye dispersing through water, we are passing through a Multiverse that can be of any proportions, but are in direct contact with 20 times our portion of cosmos.

This entire 'bike on top of a train' model agrees with the H_0' model in the end result, but why? The answer doesn't lie in our cosmos, but the deceleration of the cosmos in which we are contained. The harder the 'train brakes,' our clock, as measured by looking at the top of the train, speeds up. This rate of change is slave to G' according to the equation set:

$$tp' \rightleftharpoons \sqrt{\frac{hG'}{2\pi c^5}}$$

$$t' \rightleftharpoons t_0 / \sqrt{1 - \frac{2G'M}{rc^2}}$$

$$H'_0 \rightleftharpoons \sqrt{\frac{8\pi G'}{3}p - \frac{kc^2}{a^2} + \frac{\Lambda c^2}{3}}$$

And given by the slope:

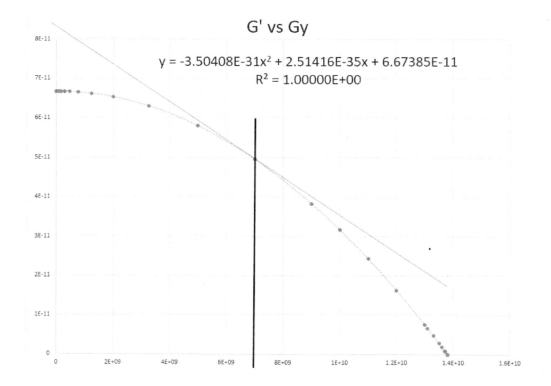

And slave to the relationship:

$$\frac{RL_0}{SL_0} = \frac{RC_0}{SC_0}$$

Where

$$R_{observed} - R_{expected} = RL_0$$

That larger system (which qualifies as the proof of the Multiple Universes Theories) holds the key to the properties of our cosmos, including most particularly gravitation, and the mutable nature of these constants and their sources, G', H_0', Lp', and tp'.

In addition, the considerations of this *subspace structure* have to be taken into account; from my observations over half a century of studying the subject at hand it appears that at least *most of* the 'paradoxes' littering physics have their basis in observing a phenomenon from two simultaneous perspectives. The *subspace structure* (the larger system that contains our physical cosmos) is the most obvious, detected, observed, ignored or miscomprehended candidate for the missing perspective vantage point.

And then we have as an over simplified example, one variation on critical density is given as

$$\rho_c = \frac{3H_0^2}{8\pi G}$$

Rearranging this gives

$$H_0 = \sqrt{(\rho_c 8\pi G)/3}$$

Since H_0 is dependent on t' as recessional velocity:

$$t' = \frac{t_0}{\sqrt{1-(\frac{v}{c})^2}}$$

$$G' = 6.67384(80) \times 10^{-11} \text{ m}^3/\text{Kg } (t')^2$$

Then we can write in fractal form:

$$H_0 \risingdotseq \sqrt{(\rho_c 8\pi G')/3}$$

If the value H_0 is interdependent on G', then H_0 turns out to be an artifact of G', and the renormalization can only result in a value of exactly 1, with respect to the critical density of the universe:

$$\Omega \equiv \frac{\rho}{\rho_c} = \frac{8\pi G \rho}{3H^2}$$

Given that G' and H_0 in this equation are a fractal, this resolves the issue of the very exacting critical density issue that has gone unresolved for half of a century.

Again, these discussions of the large scale structure of the universe are for edification and also necessary to clear up huge misconceptions regarding what space-time is and what its structure is on a large scale.

PLANCK SCALE CORRECTIONS TO THE ALCUBIERRE SPACE-TIME MANIFOLD

PI, IRRATIONAL VALUES; UNCERTAINTY, THE MECHANISM THAT ALLOWS ONE PLANCK INTERVAL TO REMAIN IN AN ISOLATED SPACE-TIME DOMAIN, YET PROCEED TO THE NEXT PI

This section is critical in understanding how one Planck interval of time proceeds, in nature, to the next Planck interval of time, providing continuity, referred to as the 'Planck Flow.' This is sometimes referred to as 'unitary time' in discussions regarding the Quantum Zeno and Quantum Anti-Zeno Effects. It is also a mystery in Quantum Mechanics regarding the mechanism how one Planck interval of time (or space) proceeds from one to the next. To date, no mechanism has been proposed.

If we take a look at the value 'pi' and its presence in the equations regarding the Planck intervals of length and time:

$$Lp = \sqrt{\frac{hG}{2\pi c^3}}$$

$$tp = \sqrt{\frac{hG}{2\pi c^5}}$$

The value 'pi' is in itself an irrational number with no discrete value. As approached from either side it appears more as an asymptote as one defines the number of decimal places, to this day may be considered infinite:

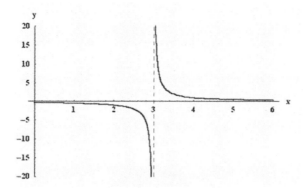

With respect to the number of decimal places of the irrational value 'pi.'

$$\lim_{x \to \pi-} f(x) = \frac{1}{0-} = -\infty \text{ and } \lim_{x \to \pi+} f(x) = \frac{1}{0+} = +\infty$$

The Planck unit of length and of time can therefore have no 'discrete' value. It is not a matter of being on a quantum scale. The values Lp and tp can be as large as a football field and still the values can have no discrete 'edge.' My 'quantized meter stick' by which I measure phenomenon also can have no 'discrete' value, regardless of scale.

It is easier to look at the Planck length for the argument. The value Lp^2 cannot have any meaning in ordinary space-time. For instance, the value Lp^2 defines a square of two right triangles with a hypotenuse of:

$$h = \sqrt{Lp1^2 + Lp2^2}$$

Where, here, *h* refers to the hypotenuse of the triangle, not to be confused with Planck's constant.

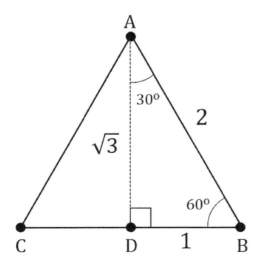

Where h refers to the hypotenuse, which in this case ends up being the square root of three. The entire equilateral triangle approach looks deceivingly appropriate along the outer edge of three equal Planck lengths, but the details of its structure are an impossibility in normal space-time. It must obey the most basic Pythagorean Theorem; else, it fails to exist. Since the triangle's hypotenuse cannot exist in normal space-time, the triangle cannot exist in normal space-time. We find this problem persists with every geometric shape known, non-integer values of Lp within the geometry that make any geometric shape in normal space-time impossible on a Planck scale.

I start with a triangle because this is the favored Boltzman solution to Holographic Theory, S=A/4:

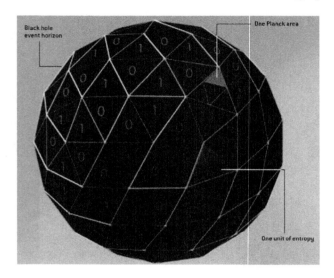

Since it is irrefutable that a triangle cannot exist on a Planck scale, then it is irrefutable that the solution S=A/4 cannot possibly be correct. Also, as a solution to the Information Paradox it is irrefutably incorrect. As I have stated, *information* cannot possibly reach the Schwarzschild radius but can only asymptotically approach it for infinity. There is no paradox in this, no loss if *information*. This also dismisses the entire mechanism of Hawking Radiation which requires the infall of one of a particle-antiparticle pair, which cannot happen due to asymptotic time dilation.

In addition, h is not an integer value of Lp. Therefore, a right triangle, an equilateral triangle, or any of the basic shapes used to describe the Holographic Principle of Quantum Mechanics is impossible for this reason. Each leg of the length is given by pi, an indiscrete value with an infinite number of decimal places describing no endpoint, and therefore does not exist in normal space-time (the length of the leg of the triangle). Therefore, neither the outer edge, which we had the greatest confidence in, nor the internal structure, namely the Pythagorean demands of the hypotenuse, are possible using Planck values or principles in normal space-time. Again, using encapsulated dimensions will not resolve the issue, as these encapsulated dimensions fall below the domain where they can resolve the Pythagorean demands, and for that matter, the indiscrete endpoints of the Planck lengths themselves as a result of being a multiple of pi, an irrational number. Compactification of a set of dimensions within a length that has no discrete endpoint cannot resolve the issue.

The definition for the Planck interval needs correction. Furthermore, as we have been investigating, G' is a floating value, and that seems to fractalize with tp' and Lp'. These things need to be taken into consideration.

The engine is dependent upon correct definitions for Lp' and tp', as well as G' in order to function, as these three values define the value and shape of space-time, and as such the artificial reshaping and revaluing of space-time. They have to be correct. Leaving such huge gaps, as is obvious using 6th grade arithmetic is not an option.

The same principle holds true for a circle, box, sphere, or any two or three-dimensional 'shape' in ordinary space-time. All have values that are not integer values of the Planck length and therefore not possible in normal space-time. On a Planck scale shape is a non-sequitur.

If we take a look at the ancient Greek process that was first used to establish the value pi called 'squaring the circle.' Using this method, Archimedes (circa 250 BCE) showed that the value of pi lay between 3 + 1/7 (approximately 3.1429) and 3 + 10/71 (approximately 3.1408). Again, using this method in 1914 with a ruler and compass, Srinivasa Ramanujan gave geometric constructions for:

$$\left(9^2 + \frac{19^2}{22}\right)^{1/4} = \sqrt[4]{\frac{2143}{22}} = 3.1415926525826461252\ldots$$

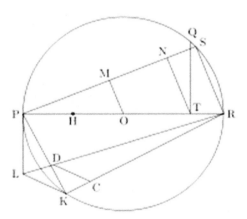

The lengths PK and SR are indiscrete values. If we look at the method for computing pi, what pi is, we find that it has a limit at the Planck length, with respect to the lengths PK and SR. There is only so far you can square that circle until your values PK and SR are 1LP. This is entirely arbitrary, dependent on the diameter of the circle, which can be of any magnitude, even hypothetically larger than the cosmos. If this is the case, and our circle can approach infinite diameter, than our values PK and SR can extend out to an infinite number of decimal places. Therefore, pi remains a truly transcendental number, if and only if given the circle can approach infinite diameter in some infinite domain. If the circle is finite, then PK and SR are limited to Lp, and the value pi, *in this cosmos* is not transcendental, but has a boundary at Lp.

That is clearly, in this cosmos, the circle cannot be of infinite diameter, thus the limit of PK and SR are Lp, and pi is not transcendental. If and only if pi is a function of some infinite domain, which must exist, can pi be transcendental.

Computer algorithms modeled by mathematicians who do not think about such things ignore the model and only crunch numbers. The model dictates that pi can have infinite decimal places if and only if our circle is of infinite diameter in an infinite domain. Since our domain is finite, has a lower boundary of the Big

Bang and an upper boundary of the present, no such model is possible, only hypothetical. Thus, pi is transcendental only as a hypothetical case, not as a real model. Furthermore, our finite domain is changing in size constantly, and we have no idea what the actual limit is, therefore, any true calculation of the value pi is non-sequitur.

In any case, the value 'pi' leaves us with an indiscrete value for the Planck units of both length and time with no definite 'edge' and is not a matter of scale but the result of being based on an irrational and transcendental value 'pi.'

Since there can be no defined 'edge' or discrete value for the Planck units of length and of time then the 'fuzziness' of these values which cannot be approached from either negative infinity or positive infinity means that there is simultaneously both an overlap for the Planck units of length and of time and also a 'gap' between them:

Gap:

Overlap:

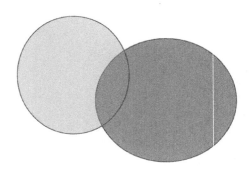

Here, of course, the overlap and gap are greatly exaggerated. However, the transcendental nature of the value pi brings about a state, on a quantum scale, of uncertainty, as will be explained shortly.

The progression from one Planck interval to the next then is the result of this simultaneous overlap and gap.

Space-time, then, is far less quantized and discrete, regardless of scale, then previously considered. Moreover, the progression of one Planck interval to the next is likely the result of this simultaneous overlap and gap between intervals of Planck Length and Planck Time. This fuzzy nature, this simultaneous overlap and gap of the Planck unit of time and of space allows, *in my opinion,* for the progression, yet isolation of one Planck interval to the next.

That is, the fundamental principle that allows for the Planck flow of one Planck interval of space-time to the next is this simultaneous overlap and gap, the indiscreet nature afforded by the value, pi.

The idea that at a quantum scale things become 'pixelated,' based on the Quantum Foam, the impossibility of shape on a 2-D or 3-D scaffold, and this fuzzy or indiscrete nature where Planck intervals are simultaneously discrete and overlaid makes the concept of pixilation an absurdity. Again, the urban myth will be repeated so many times that eventually young physicists will come to *believe it* as Orthodox Canon with complete disregard for the absurd nature of it and eventually go as far as to summarily dismiss any statement otherwise. It will be written into graduate level texts, and so on. The word 'pixelation' comes from uneducated computer geeks who think they know quantum mechanics because they look words up on google.

In a universe where the Planck unit of time is truly isolated with no overlap, this in turn requires, for the progression of common time, the annihilation and restoration of each and every thing down to the quarks and whatever presumably makes up the quarks 10^{43} times per second. This is because in such an isolated system, there is no way to 'carry' such *information* from one Planck interval to the next, and no satisfactory mechanism has ever been proposed to account for this. In fact, the question is never approached. One can possess all the momentum in the world and it is going nowhere because the next Planck interval of time does not yet exist. When the next Planck interval of time exists, the current Planck interval of time has ceased to exist. If the two exist to any degree at the same moment there is overlap, and that is exactly what I am describing here. This overlap is caused by the irrational value pi, which causes both simultaneous overlap and isolation because of its indiscrete value; it lacks an endpoint, and allows *information* to be transferred from one Planck interval to the next in both space and time.

I explained the Planck Flow as a 'standing interval' with each interval ceasing to exist and the *information* carried on as a photon or photons. This quantum jump (the Planck Flow of one interval to the next) can only occur at v=c or v=0, likely therefore v=c. The endpoint has no definitive destination in space-time. That is, suggesting its only possible location at v=c is 1Lp away is not necessarily the case. The only limitation is that Lp is defined by *c* and by pi. In our cosmos pi cannot be a 'transcendental number' because our cosmos is not an infinite domain.

That leaves Lp a function of subspace.

The next obvious question is, 'how much overlap can we be talking about in a transcendental value that goes out to an infinite number of decimal places?' First, we do not need to go out nearly that far in decimal places to overlap, the Planck scale is only 10^{-35} meters.

With respect to the substructure of space-time in quantum dynamics referred to as the 'space-time foam,' this feature of both Lp and tp being based on an irrational and transcendental value with no discrete edge, a characteristic previously not considered, gives the quality of space-time a much less 'pixilated' nature. In fact, the modern terminologies trying to render a description of the quantum scale structure of space-time taking on a pixilated characteristic slip hopelessly far away from anything that can possibly resemble the truth. The space-time foam, the simultaneous overlap and gap, the shapeless, formless nature of the Planck interval, made up of indiscrete, irrational values carrying *information* from one Planck interval to the next make this approach at a 'pixilated' characteristic impossible, and in fact exactly opposite of what is correct. On a quantum scale, space-time takes on Wheeler's foamy characteristic, highly dynamic, churning, formless, shapeless, and even timeless. 'Pixilation' requires among the most impossible shapes of the quantum scale:

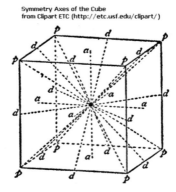

Symmetry Axes of the Cube
from Clipart ETC (http://etc.usf.edu/clipart/)

This represents a 'pixilated' uniform shape as proposed for the quantum scale. It is laced with impossible attributes and geometric properties that are impossible on a Planck scale because of the impossibility to dice the Planck length into so many non-integer values of the value Lp. All hopes of discussing space-time in terms of being somehow 'pixilated,' and I use this in terms of being in concert with Wheeler's space-time foam, are so far from being in any way possible it is amazing that the idea has even been given consideration whatsoever in the face of the obvious.

The quality of the substructure of space-time then becomes smoother than this 'pixilated' vision as a result of the simultaneous overlap and gap between Planck intervals, then given the highly dynamic foamy characteristic, but not totally 'smooth' by any means. Perhaps as 'smooth' as 'The Perfect Storm.'

This particular quality of the indiscrete values for the Planck units of length and time allow for the progression of unitary time and reduce the quality of the substructure of space from a hardened 'pixilated' quality to a more 'smooth' characteristic, but again, by no means totally 'smooth.' It is more like a storm.

And again, this indiscrete, irrational value for the Planck units of length and time allow for simultaneous overlap and gap, both allowing for the progression of unitary time and also simultaneous isolation of Planck intervals of space-time such that no two Planck intervals have a common present, yet exist as a 'standing interval.' The *interior* of the Schwarzschild surface exists as photons, the *information* of prior Planck intervals.

This is Zeno's arrow in flight. A single snapshot. No snapshot exists before it, it has ceased to exist. No snapshot exists, yet, of it in the next frame of where it will be in the future. The QZE isolates this snapshot, in the Planck Flow, and keeps it from progressing to the next frame. The mechanism is constant observation. Why constant observation works is hypothesized by myself as *information* falling into an infinite domain (the observer). We see this at the event horizon of a Black Hole, for instance. Observation, within the parameters of Von Neumann arguments (given a quality definition for consciousness) refers to *consciousness* being an infinite domain. Mechanisms carry out the work for us as extensions of our selves.

In order for *information* to be transferred between discrete Planck intervals, this would require a mechanism that occurs outside of normal space-time, because all of normal space-time is described by discrete Planck intervals. This hypothetical mechanism would have to transfer *information* between discrete Planck intervals outside of normal space-time. Furthermore, this hypothetical mechanism would have to operate in a dynamic system of quantum foam between shapeless forms of Planck intervals. This hypothetical mechanism would have to be capable of transferring every plausible form of *information,* mass-energy, momentum, velocity, and so on.

In this case, Leonard Susskind's -1 Law of Thermodynamics, '*Information* cannot be destroyed,' takes precedence over the choice of mechanisms available for hypothesis, and it is fortunate that we have this limiting factor as a guide-stone. Considering each Planck interval to be 'unitary,' that is, absolutely discrete, this would require complete annihilation of all *information at the termination of a Planck interval of time* and then complete reconstruction of exactly the same *information* at the beginning of the next Planck interval. This sounds like magic because it is. It is not mysterious it is absurd. In this hypothetical scenario, we have not even fashioned a hypothesis regarding how and why the next Planck interval begins when such a description, according to all of the laws of thermodynamics suggests everything in this cosmos should come to an abrupt end.

Keep in mind that it is permissible within the -1 Law of Thermodynamics that *information* can be reduced to some chaotic form. This is another failure in Holography, Chaos Theory, rather, the lack of it. I define *information* as organized information that represents a tangible thing, process, etc. Holography has not dealt with the issue of organized information nor tangibility or any time dependent process, *which is all of them, every force of nature.* That is, Holography has dismissed every force of nature rather than deal with the complexity of it.

Later, in Temporal Mechanics 101 I provide a solution to Holography that does not have all of the problems I have listed thus far. It's honestly not that difficult.

The simplest answer that resolves both the transfer of *information* from one Planck interval to the next and provides a hypothesis as to *why* the next Planck interval of time forms in the first place is the indiscrete nature of the Planck interval based on the transcendental nature of the value pi. That is the simplest explanation, and from there it gets more complicated, and in turn more absurd.

The fact that the endpoints of the Planck interval are indiscrete is not conjecture, they are founded on the value pi, and the indiscrete nature of their nebulous endpoints is therefore *irrefutable* in any case. That is, rather than put forth a hypothesis requiring the indiscrete nature of the Planck interval based on the transcendental nature of the value pi; the Planck interval is indiscrete based on the transcendental nature of the value pi regardless of our need to support a hypothesis. That is *irrefutable.*

What causes the progression of the Planck Flow is the *information* contained within the Planck volume contains time as a fundamental.

In this system, no such hypothetical mechanism of information transfer outside of the normal space-time domain is required. Planck intervals overlap due to the irrational, transcendental value pi, and remain discrete (standing interval), simultaneously, for the same reason. In this case, the mechanism for transfer of *information* from one Planck interval to the next is simple overlap, or being *indiscrete*. The indiscrete nature of the Planck interval is the result of being founded upon an irrational number, pi, extending out to perhaps an infinite number of decimal places, meaning that it can in all actuality never be placed anywhere in normal space-time altogether. Thus, not only does the Planck interval have no discrete endpoint in normal space-time, but also it has no discrete location in normal space or normal time altogether as a result of its boundaries being non-finite. That is, pi will never reach the value 3.2, the approach is an *irrational asymptote of turbulent nature,* but it lacks a discrete value and thus a discrete location or endpoint.

This, in turn leaves us with the further evidence for the Multiverse solution of an infinite substructure of space-time, else pi would stop at the Planck limit.

THE DEVICE, IN GENERAL TERMS

The basic principles are: use the Quantum Zeno Effect and Quantum Anti-Zeno Effect (detailed later) to alter the progression of unitary time (the Planck Flow). From that point the demands of General Relativity will produce a predictable reshaping of space. The exact control of the QZE and QAZE via the equations I will provide here will 'paint' an Alcubierre Space-time Manifold that is used as a superpositioning mechanism. Velocity is probably not the correct term to apply.

In this engine design, the spin states of entangled particle-antiparticle pairs in large populations is chosen as the measured phenomenon because there is no argument that allows for a particle's spin state to be altered as a result of the rapidity of the measurement. That is, in Quantum Mechanics, the 'Measurement Problem' often comes into play, such as we see as common with arguments regarding the double-slit experiment, where the mechanism of measurement comes into question as affecting the result. Measuring quantum spin states does not have this 'Measurement Problem' characteristic, until one derives Quantum Entanglement states, the EPR paradox, Bell's Theorem's, and so on. However, the noted paradoxes are not necessary to produce a numeric and quantifiable real time result with respect purely to spin state; the paradoxical nature of these noted issues only pertain to *prediction of information, not the real time measurement or result of such information.* In fact, the EPR paradox, Bell's Theorems, and so on are arguments that are reverse engineered from such data in an attempt to explain how the results were obtained. They have no effect on the data a priori.

In the absence of the esoteric nature of the a priori arguments regarding the paradoxical nature of how the data will appear, the quantum spin states of entangled pairs of electron-positron pairs are measured in large populations at very high speed in order to produce a Quantum Zeno Effect (and Anti-Zeno Effect) on a macroscopic scale. This also omits the possibility that the QZE is an effect caused by some other phenomenon than measurement rate as has been suggested in cases regarding nuclear decay and so on. The engine is pure QZE. The QZE affects the flow and progression of linear time, t'. The *demands of General Relativity require an alteration in the shape of space.*

This inter-relationship between t' and the shape of space are common all throughout nature and pervade every corner of the cosmos. Producing the QZE to obtain the same effect is the next evolutionary step in human thinking in the artificial control of the shape of space-time. By controlling the rate and/or degree of the QZE in a macroscopic region of space by way of altering the rate of detection and measurement, the shape of space-time is carefully manipulated to produce an Alcubierre Space-time Manifold, exactly as Alcubierre described without much modification; other than to correct for Planck scale aberrations and to flip the manifold to account for Lorentz inverted equation for 'length contraction.'

To date, there have been some issues regarding radioactive decay as the measured subject and questions or debates concerning various alternative explanations negating the QZE as a problem with the means of measurement. We circumvent all such debates (that I have seen so far) by measuring spin states of particle-antiparticle pairs. I believe Raizen put this away by producing both a QZE and Anti-QZE (the relative speeding of linear time, e.g., t'<0) by monitoring quantum-tunneling electrons. Nonetheless, radioactive decay is inappropriate for this engine design due to several factors and monitoring of electron-positron pair spin states will be the method of choice.

In exacting terms of General Relativity, if time slows (dilates), classical approaches have space curving inward like approaching a massive object such as a planet, star, or black hole. If times speeds up, space curves 'outward,' for lack of a better term. This 'outward curvature,' or as I like to say, space-time inversion hasn't been considered feasible; considered to require 'exotic matter,' 'negative energy,' and other exotic conditions, confirmed to occur in nature by LIGO as gravitational waves as a result of two coalescing Black Holes as depicted here in this supercomputer simulation (shown below). The phenomenon in question, as noted prior in this paper, is depicted as the upward spires; the two Black Holes are out of sight in the two distinct sharp wells on either side. No exotic conditions are required to produce the Alcubierre Space-time Manifold, the elusive upward spires occur in nature under extreme, but not exotic conditions.

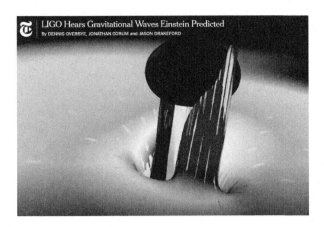

Rather than referring to the curvature of space as being 'inward' or 'outward,' which is ambiguous, I prefer to refer directly to the Planck unit of length, L_p, in this case an alteration of the Planck unit of length, L_p'. In any case, we need to avoid the semantics of what is happening to the curvature of space, what is observed, and what is 'real.'

Oddly, at various conferences when physicists described the need for 'exotic energy' to produce the upward spires; *half of them pointed at one end of the manifold and half at the other.* Clearly, the Alcubierre Space-time Manifold is not understood even at the grass roots level. Requiring a chemical rocket 'boost' because its symmetry didn't know which way to go and actually getting funded for it was more than I could bare.

We need to measure these values directly. There is currently no precedence for this process, and no mathematical foundation. There are numerous phenomenon which must be measured, numerous approaches to the measurement, and numerous constants and other values that have to be determined, for the first time, experimentally. These cannot at this time be predicted, as there is, again, no precedence for this technology.

I will state here that after years of conducting research into the QZE it appears obvious to me that each type of system, that is, one type of nuclear decay under a certain set of conditions vs. another type of nuclear decay under the same or different set of conditions has an entirely different set of constants that lead to the end result of time variation. This is why the clarity of any such set of constants has remained elusive up until now. That is, given a bit of cesium monitored via a set of conditions, mass of cesium, distance to detector, environmental conditions, and so on, any measurement rate, R, and its effect on the progression of unitary time, the QZE, will be unique to that system and not universal to all QZE phenomenon. Furthermore, there has been no attempt to catalog such data. Eventually a large enough population of QZE data under carefully manipulated experimental conditions will exist such that these values can be derived, but that is a long way off. This engine will have a unique set of properties determined via experimentation (R&D), the next generation of engine will be improved upon by observations of the prototype, and so on.

The reason for this, in the barium decay example, is that each variable; the decay type and rate of cesium nuclei to barium-137 via beta decay or gamma via barium-137m, given 85 neutrons and 55 protons, the 85 neutrons vying for the probabilistic weak decay route of internal quark-quark color exchange and consequent emission of a W boson whose subsequent decay is also probabilistic as to the distance and type, cutting through the Z barrier of multiple electron orbitals before we even leave the atom are all time dependent. The distance to the detector, the environmental conditions between the atom and the detector; is the atom shielded by other cesium atoms, is the entire system enclosed in a vacuum, an inert gas, what is the system's elevation in Earth gravity well, is the system near the equator where its linear velocity is greater than at the poles, and so on… Even rudimentary questions such as, 'how is the cesium sample mounted,' and such are relevant to the effect of the progression of time in the system. One type of mounting can incite quantum-tunneling phenomenon where another may inhibit it. Quantum-tunneling can be an issue with beta decay. Why would a sample quantum-tunnel over a present barrier? Exactly for the same reason a man climbs a mountain, because it's there. In fact, as experiments would suggest, not the least of which would be the famed double-slit experiment, providing such a barrier would cattle-heard the beta particles in all the wrong places, thus mounting of the cesium sample is a major issue in a QZE experimental setup. Every little thing where one can measure tp and/or Lp introduces variables in the QZE, and thus no two QZE experimental setups are identical. At the very least, the overall configurations are unique and the QZE constants all have to be derived experimentally and cannot be predicted, at least not at this primitive stage of our experience with the phenomenon with the QZE. This is the reason no one has derived or noticed any consistency between QZE experiments, albeit, any particular experimental setup, if exactly reproduced, has reproducible results. This has become the hallmark of quantum computing, and a vital component in keeping quanta alive long enough to perform vital calculations. Quanta are actually kept alive in a quantum computer via the QZE in order for the system to function. It has thus become an industrial process.

In order to artificially alter the shape of space, we slow apparent time by taking more rapid measurements (I am using a classic QZE argument here, not approaching the QAZE at this time), time dilates from our perspective, and space-time *must* therefore shape itself accordingly else violate General Relativity. As the progression of time 'apparently' slows, t' increases, G' decreases, resulting on both Lp' and tp' to diminish. The information must therefore take a longer path to the detector, crossing a greater number of Planck units of both length and time, resulting in an 'apparent' slowing of time of the observed system. (If there is one paragraph you take home with you from this text, this is the one).

That is, we review once again the equations for time dilation, t' and tp', length dilation, L' and Lp', and G':

$$G' = 6.67384(80) \times 10^{-11} \text{ m}^3/\text{Kg (t')}^2$$

$$t' = t_0 / \sqrt{1 - \frac{2GM}{rc^2}}$$

$$\pm t' = \frac{t^0}{\sqrt{1 - (\frac{v}{c})^2}}$$

$$t'_p = \sqrt{\frac{hG'}{2\pi c^5}}$$

$$l' = l_0 / \sqrt{1 - \frac{2GM}{rc^2}}$$

$$l' = \frac{l_0}{\sqrt{1 - (\frac{v}{c})^2}}$$

$$Lp' = \sqrt{\frac{hG'}{2\pi c^3}}$$

Note here that I am now using L' in the non-conventional sense according to the proofs provided.

In a classic QZE experiment, we take more rapid measurements of a regular phenomenon, such as in our case, a beam of electron-positron pairs. We are filtering off the electrons, sending them through a Stern-Gerlach mechanism, and using a conventional detector to differentiate their spin states by their position on the detector screen, just for argument sake at this time. As we increase our detection and measurement rate, time begins to slow for the system. This slowing of time is indifferentiable from time dilation in every way. There is no change in preferential perspective that can yield any useful information whether or not this is indeed a relativistic effect of some sort, as a change in preferential perspective will alter our 'tick rate' for all observable phenomenon outside of our frame of reference, everywhere. We go to measure our QZE system with our meter stick that is quantized within our frame of reference and can only determine that the QZE result is a quantized factor. There is nothing we can do in order to determine that the QZE is differentiable from a relativistic-like effect. Therefore, all of the equations of a relativistic nature apply.

As time slows, t' in equation 4 increases. As t' in equation 4 increases G' decreases. As G' decreases, tp' in equation 3 and Lp' in equation 7 decrease. The *information* thus has to cross a greater number of Lp and tp, Planck units, in order to reach us; that is, the *information* has taken a *longer path to reach us. We have artificially reshaped space-time via the QZE. The longer path is equivalent to a slowing of the progression of time, or a Ricci curvature of space-time. In every case, the QZE we have produced is behaving indifferentiably from both Special and General Relativity, simultaneously.*

Furthermore, Mark Raizen has produced the Quantum Anti-Zeno Effect (QAZE), the *speeding* of the progression of time by carefully altering the data acquisition rate of a phenomenon (he used electron tunneling). In this case; as time speeds, t' in equation 4 decreases. As t' in equation 4 decreases G' increases. As G' increases, tp' in equation 3 and Lp' in equation 7 increase. The *information* thus has to cross a lesser number of Lp and tp, Planck units, in order to reach us; that is, the *information* has taken a *shorter path to reach us. We have artificially reshaped space-time via the QZE, opposite the former scenario. We have in this case produced a space-time inversion, as shown in the LIGO model of two coalescing Black Holes and confirmed by LIGO data to have actually occurred in nature, NOT under exotic conditions.* Raizen produced the effect on a microscopic scale.

We have just described, the artificial control of the shape of space-time both inward and outward, sufficient to produce an Alcubierre Space-time Manifold (we'll get to the specifics a bit later on). Each point on the manifold depicted below is 'painted' in thus fashion, according to the data detection and measurement speed from a large population of quantum events according to their measurement rate, R.

In producing the Alcubierre Space-time Manifold, the measurement rate, R, of the quantum spin states, {+1/2,-1/2} determines the shape, height, width, curvature, and so on of the upward spire (Quantum Anti-Zeno Effect) or downward spire (Quantum Zeno Effect) and the flatness of the isolated space-time region at the center of the manifold.

A good visualization of the Alcubierre Metric is depicted as: (again)

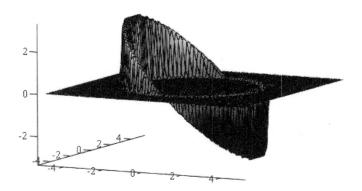

This visualization (albeit gratuitous) is extremely useful in order to grasp the way space-time must be shaped in order to 'paint' an Alcubierre Manifold by adjusting the flow of time such that, as depicted here, time slows along one path of the manifold, increasing Lp', and time accelerates faster than 'normal' space-time on the opposite slope, decreasing Lp', and remains 'flat' in the domain in the center (referred to as the 'top hat' region) of the phenomenon. The depicted upward spire had physicists in a state of confusion for decades. However, as I have pointed out, this upward spire has been proven to exist in nature, observed now by the LIGO gravitational wave observatory, and produced on a microscopic scale in Mark Raizen's laboratory [M. C. Fischer, B. Gutí errez-Medina, and M. G. Raizen, Department of Physics, The University of Texas at Austin, Austin, Texas 78712-1081 (February 1, 2008)]. The conditions that caused it (the large space-time inversion) were extreme, the coalescing of two moderate sized Black Holes, but not exotic, as in 'negative energy' or 'exotic matter,' and so on, conditions that remain yet to be defined. The space-time inversion will not require two coalescing Black Holes to produce, of course. The space-time inversion is produced by detection and measurement, only. To this day, although the depiction is out there, I do not think that the physicists involved have put it together that this upward spire in the microsecond spark that fronted the gravitational wave is the same phenomenon sought after for decades in the making of this space-time manifold. Too much focus, not enough generalization.

Here, the bubble represents a population of electron-positron pairs being produced and emitted at very high luminosity. We imagine this region as 'the engine.' The electrons and positrons are of course going off in opposite directions, but the bubble is for illustrative purposes. We can consider the bubble a plasma floating in a magnetic field.

In order to slow the progression of 'unitary time,' we take extremely rapid measurements of 'a system' along the outer edge of that bubble. The more rapid the measurements, the greater the factor of slowing time, the more space curves. I use the mentioned 'Splitting Point,' designated SP, explained a little further on, to determine the measurement rate, R, in order to produce the Quantum Zeno and Quantum Anti-Zeno Effects to 'paint' an exacting space-time manifold as originally proposed by Alcubierre, depicted at the center of the bubble. It is by measuring the phenomenon on the surface of the bubble that the space-time manifold at the center of the bubble manifests.

That is, the sphere of electron positron pairs is reshaped into Alcubierre's Space-time Manifold via the QZE and QAZE by detection and measurement. Rather, it is more accurate to say that the Manifold reshapes the sphere, which actually remains a sphere, but the shape of space-time alters. This is exactly analogous to light 'bending' around massive objects (gravitational lensing). Light doesn't actually 'bend,' it follows the path of curved space-time.

As we control the measurement rate along the surface of the bubble in a predictable fashion. This predictable fashion will come as a result of determining the constants unique to this system of QZE and QAZE and governing our measurement rates along the surface of the bubble, which is a high population of Quantum phenomenon, quantum spin states of elementary particles.

The actual coefficients of time *dilation* or more precisely, the progression of time, since both dilation and anti-dilation effects must be generated, would have to be determined experimentally. There is no provision in Special or General Relativity for 'anti-dilation.' However, there is provision for anti-dilation if the observer's perspective has a t' sufficient to allow for a t_0 of the observed system that falls below t' for the observer. That is, anti-dilation doesn't necessarily suggest t'<0, it merely suggests t'<t_0 for the observer's frame of reference. It's not very complicated. Stick three guys at three points falling inward toward an event horizon, imagining that they can see each other's clocks. The guy in the middle has a unique t' that is >t_0 for one observer but <t_0 for the other. Nonetheless, there is no law of physics or thermodynamics that forbids a t'<0 in any case. That is, in Orthodox Quantum Mechanics, the fastest tick rate is that which occurs in the flattest space-time, but there is no *law* in QM that forbids a faster tick rate, but there is plenty of *opinion*.

As an example of t_0 varying for each observer in a gravity well, for instance looks like this:

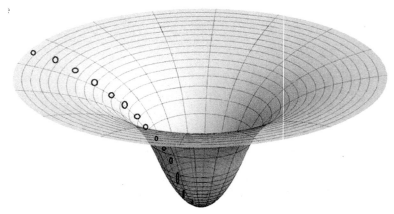

Each observer, designated by a marker, has a t_0 value differing from each other observer. Those deeper in the gravity well see t' as greater than zero, or t'>0. The Quantum Anti-Zeno Effect does not break any rules in this fashion. In this particular example, the marker at the outermost edge has the fastest tick rate. Thus, for the observer deepest in the gravity well, he sees that marker as t'<0. Less than zero does not mean *negative,* or backward in time, it means that the observer deep in the gravity well sees the temporal marker on the outermost edge as 'going faster,' according to his clock. In fact, the observer deepest in the gravity well sees all of the other markers as progressively going faster according to his clock.

Since we do not know what the actual tick rate of the local conditions in our vicinity within the cosmos are, and that includes time from the Big Bang until now as well as locality in the entire, not just visible cosmos, we do not know what t_0 is. As I showed on the graph some way back, both time, G', H_0, and a host of other variables have been constantly changing according to this second order curve I presented (pg 72).

This is a generic term that indicates a 'quickening of the progression of time.' In the simplest sense, a GPS satellite in orbit experiences this 'quickening' relative to a clock on the Earth's surface (and also a slowing to its orbital velocity). However, the Lorentz transformation doesn't actually provide a result of this sort directly. An approximation of the time coefficients I defined in QTD (A separate text: *Temporal Mechanics 101*) as being approximated by red-shift as perceived by a stationary observer, keeping in mind that in QTD red-shift is quantized according to:

$$f_{observed} = f_{emitted} \sqrt{\frac{1-((nLp/xtp)/(1Lp/1tp))}{1+((nLp/xtp)/(1Lp/1tp))}}$$

Meaning that it is probably correct to assume our Alcubierre Manifold will take a 'quantized' form, rather than a smooth metric as the orthodoxy depicts, which also serves as a gross approximation. However, this equation is in terms of velocity, noting that (nLp/xtp)/(1Lp/1tp) is equivalent to stating v/c, and is unambiguously quantized. *Smooth motion in quantized space-time is not possible. Smooth acceleration in quantized space-time is also not possible. Since acceleration and gravitation are relativistically equal,*

General Relativity is quantized. Since (nLp/xtp)/(1Lp/1tp) is equivalent to stating v/c and smooth motion is not possible in quantized space-time, Special Relativity is quantized. This takes us back to our frame of reference, where our meter stick is quantized within our frame of reference and it is impossible to measure a non-quantized value for an object at high velocity (Special Relativity) or deep within a gravity well (General Relativity) because of the nature of the meter stick that is quantized within our local frame of reference.

Now we run into the case where space-time itself is quantized and motion and acceleration must take on quantized forms as described prior in this paper with only two options of motion possible, v=c and v=0. Now we have the case where the relationship (nLp/xtp) yields the quantized value for L', Lp', t', tp', and G'.

First reviewing our general principles of quantized motion:

At v = 0.5c, we are faced with

1. go ½ Lp in 1 t_p

That is not possible because this requires a structure finer than Lp (a Planck length) will allow:

Figure 1

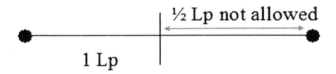

2. Or, go 1Lp in 2 t_p:

   Since proceeding at >1Lp/ t_p is exceeding light speed, this is forbidden.

   Since going < 1Lp is forbidden because it requires a structure finer than Lp will allow.

We are faced with motion taking on the following characteristic:

Go at v=c for 1 t_p, stop for 1t_p, go at v=c for 1 Lp, stop for 1 t_p; etc.,

interval	v	behavior	t'	L'	m'
tp1	c	go	infinite	0	infinite
tp2	0	stop	t0	L0	m0
tp3	c	go	infinite	0	infinite
tp4	0	stop	t0	L0	m0
tp5	c	go	infinite	0	infinite
tp6	0	stop	t0	L0	m0
tp7	c	go	infinite	0	infinite
tp8	0	stop	t0	L0	m0
tp9	c	go	infinite	0	infinite
tp10	0	stop	t0	L0	m0

Then

$$v = \frac{nLp}{xtp}$$

$$c = \frac{1Lp}{1tp}$$

$$v/c = ((nLp/xtp)/(1Lp/1tp))$$

In Special Relativity

$$t' = \frac{t_0}{\sqrt{1 - ((nLp/xtp)/(1Lp/1tp))^2}}$$

Since our meter stick is quantized to our frame of reference and there is no other means by which to measure the phenomenon other than by the meter stick that is quantized to our frame of reference:

$$tp' = \frac{tp_0}{\sqrt{1 - ((nLp/xtp)/(1Lp/1tp))^2}}$$

In Special Relativity, now using the non-conventional approach to length dilation according to the mathematical proofs presented, the term, Lp' follows the same principle:

$$Lp' = \frac{Lp_0}{\sqrt{1 - ((nLp/xtp)/(1Lp/1tp))^2}}$$

This principle would hold true even if Lp' were presented in the conventional sense:

$$Lp' = Lp_0\sqrt{1 - ((nLp/xtp)/(1Lp/1tp))^2}$$

The basic principle is that each Planck volume of space-time or a population of Planck volumes of space-time are making leaps in the form v=0 and v=c as shown in the chart above, and no other velocity is possible. Velocity and acceleration are thus quantized to v=0 and v=c with zero acceleration between v=0 and v=c. They are literally quantum jumps in velocity. At what scale this quantum scale phenomenon becomes undetectable is unknown, and has yet to be characterized. A simple thought experiment is to try and chart the velocity of an object of one Planck volume as it attempts to cross one Planck length at any velocity less than light or greater than zero. Upon working out the details in every conceivable way one concludes that the object can only travel at v=0 or v=c. The term (nLp/xtp) defines these quantum jumps in velocity in terms of a ratio of v=0 and v=c, whose mean value turns out to be *v*. On a quantum scale, the quantum components of a system may be jumping at v=c for one Planck interval of time while holding at v=0 for 9 Planck intervals of time, and we say that *v=1/10c*. However, on a macroscopic scale, we do not see either v=0 or v=c, we see what appears to be a continuous v=1/10c. However, it should be noted that as before stated, there are 10^{60} Planck volumes filling a single neutron. To put that in perspective, there are an estimated 10^{80} atoms in the visible cosmos. It is likely that this quantum leaping characteristic is not apparent even on a nucleonic scale. Thus, our quarks would appear to move smoothly and continuously within the nucleon, and no one has a clue what a gluon does or even what it is.

All of this seeming side track is imperative to the engine mechanism. We are going to do artificially what nature does. As for the space-time inversions we saw that in the LIGOs data of two coalescing Black Holes, NOT as a result of a universe worth of 'exotic negative mass-energy.' The quantization of Lp as per that lengthy explanation is such that at the Big Bang it was some ridiculous value, and now it is such that the same number of Planck volumes (remember the number cannot change due to the -1 law of thermodynamics) has just the visible portion of the cosmos some 45 billion light-years across (in comoving distance). The engine design is going to take that quantization and artificially extend the value Lp out to any value, from 10^{-35} meters to perhaps light-years. That is what a photon in nature does, it superpositions itself in thus fashion by that means. It does not 'contract' space, distort the shape of the cosmos, as the Orthodoxy explains it. It is thusly not an 'observed effect,' any more than the exact same equations lead to *real mass increase* in a particle accelerator. It is thusly not 'double-talk,' because the equation has been upside down for a century.

This apparent continuity from a quantized Planck scale is universal. However, the line or region that divides this quantized Planck scale from that which we perceive as somehow continuous remains elusive in every case. Quantum Entanglement is a bizarre condition where a quantum jump can be any order of magnitude in scale and yet remain one quanta as a Superposition. A photon perceives its own state as Superpositioned throughout every corner of the cosmos, with all distances equal to zero. Quantum Entanglement is a shift in this perception to include the *interdependent observer*. Since the photon is in fact Superpositioned with every molecule within the brain of the observer (for those who think consciousness resides in the brain) it is no mystery that this shift in perception is possible.

Quantum Entanglement takes the form:

Where $c = L_p'/t_p'$; where each have dilated to infinity. Where ∞/∞ remains undefined, c is not violated.

Which can also be used to define a superposition.

My former argument as to the indiscreet nature of the quanta allowing for the simultaneous overlap and gap, permitting the transfer of *information* between quanta does not explain the phenomenon of quantized velocity in this current argument. There is a transfer of *information* between quanta in this process of velocity, and it must follow the same principles described via the indiscreet nature of the quanta. However, the go-stop-go nature of this transfer of *information* at v=0 and v=c follows a different set of rules.

Most arguments negate any such effects on a macroscopic scale. Nonetheless, it has never been attempted before. However, in the case of this engine we are scaling a quantum scale phenomenon up to a macroscopic, even cosmological scale. We therefore have to be attentive to the effects on both scales and examine the consequences.

We then take into account the recalculation of the Gravitational Constant, G, as designated earlier for G' as:

$$G' = 6.67384(80) \times 10^{-11} \text{ m}^3/\text{Kg } (t')^2$$

In this case, t' will be a value that has to be determined experimentally in order to determine the exact coefficient of observing a phenomenon at a given rate with a resulting effect on the progression of time.

Designating the rate of observation as R, and the resulting effect on the slowing of the progression of time, be it faster or slower, that is, designating 'normal' time as having a coefficient of 1, the general expression for the rate of observation and the resulting change in the flow of unitary time, designated t'' is given as:

$$R \, \alpha \, t''$$

Here, we begin separating the designation of time dilation, previously defined by t', from the altering of the flow of one unit time to the next, designating the phenomenon as t''. That is, we will regard the Quantum Zeno Effect specifically as altering the progression of one Planck interval of time to the next, referred to as the 'Planck Flow.' This can be considered as either A) altering the Planck length or B) changing specifically the tick rate of fixed Planck intervals. As I have gone to some extent to point out, both cases are indifferentiable.

Ultimately, they (as I have explained above) are equivalent and in most cases can be used interchangeably. However, the QZE is unique from Special and General Relativity, as a given, only in that the mechanism does not involve mass or energy. The founders of Quantum Mechanics, with the keyword being 'Mechanics,' were stuck in 19th century thinking. Everything in Quantum 'Mechanics' involved mass and energy. Even Schrodinger's wave functions predicted where particles, *tiny cannon balls of solid stuff* would be. When Quantum Field Theory came along and aside from the math being beyond even most physicists, suggesting that there was no 'stuff,' just fields and wave functions, it did not catch on. Just before Feynman died he admitted that perhaps QFT was right, there were no particles.

Thus, we are not taking the USS Enterprise, made of quadrillions of tiny cannon balls, 'solid stuff,' and hurtling it across space-time. We are taking wave functions, in a universal field, and superpositioning it.

$$R \, \alpha \, t''$$

There is no guarantee, only speculation, that this relationship is consistent regardless of the phenomenon being observed. That is, t'' may have a different coefficient for observing the quantum state of an electron than for the beta decay of a neutron. There is no good reason to state otherwise until the coefficients are derived experimentally.

In this case, our value G' will determine the actual curvature of space according to the standard Einstein-Ricci curvature that I will cut and paste here for convenience sake:

$$R_{ab} - \frac{1}{2} R \, g_{ab} = \frac{8\pi G'}{c^4} T_{ab}.$$

In this case, G is replaced by G' and c is replaced by 1Lp/1t$_p$ (quantizing motion), and T$_{ab}$, the energy-momentum tensor is derived from our observed coefficient of time as proportional to Rate of observation, R, giving a new energy-momentum tensor (being time dependent) and call it T', giving a new definition for the curvature (on the right hand side of the equation) as:

$$\frac{8\pi G'}{Lp^4 / tp^4} T'_{ab}$$

Which represents just taking the 1-dimensional measurement from point a to point b, across a table-top, for instance, and determining the effect on the speeding or slowing of unitary time, which in turn changes G', in a quantized fashion, resulting in a curvature of space as obeying the laws of General Relativity. T', in this case, is an artifact (being the energy-momentum tensor) of G' and curvature, directly proportional to the derived coefficient of time for the observed phenomenon. Also note the presence of our problematic value pi. Pi and quantization are an oxymoron. We will therefore derive pi out to that point where PK=Lp.

In this example, so far a 1-dimensional measurement of a single observed phenomenon is being described, the (for a given observation rate, R) slowing of unitary time (time with respect to the system), and the curvature of space which results from that slowing of time, noting that such curvature will be quantized.

This benchtop experiment, incidentally, is our 'proof of principle.' It has to be done with exceptional care because, given there is no precedence for this technology to date, many variables have to be explored.

If we choose to observe the quantum states of an electron-positron pair as they speed off in opposite directions, measuring the spin of each particle as being either +1/2 or –1/2, this is an ideal measurement for a multitude of reasons. First, if we measure particle 'A' we instantaneously know the quantum state of particle 'B' (definition of entangled particle-antiparticle spin pairs) without actually taking the measurement, meaning that we can at least double our data acquisition rate in a single measurement. It is important to note here that taking advantage of entangled particle-antiparticle pairs by conscious observation also takes advantage of the *seeming paradox* of the particle-antiparticle pair not being separated by real space-time; *purely as an effect of observation and measurement. We gain information about what the shape of space-time is at the opposite end of the engine in a single measurement.* Again, avoiding the arguments regarding alternate explanations for the QZE, a particle's spin state is not going to be altered by the rapidity of the measurement. Electron-positron pairs are ideal because they are easy to produce in high luminosity (large quantities) and easily directed via magnetic fields.

Also, any means of probing this particular phenomenon *may* require photons which are coupled to a magnetic field, meaning virtual photons resulting from the Near Field effects of the emitter, which 1) cannot exceed three wavelengths in duration as their defined lifetime meaning control over the span of the radius of observation; thus requiring exceeding long wavelengths and 2) control over the virtual photon's path at all times with two way communication resulting in a 'Force.' (another Near Field Effect). This is one possible and possibly superior approach to engine design in that the information is for all practical purposes 'continuous.' This has to be determined experimentally and cannot be predicted by existing models.

Probing with 'real' photons, Rf, for instance, 1) gives us little feedback regarding the particle-antiparticle pairs quantum states 2) we have no control over the distance or fate of the photons in question, but have to design the architecture of the engine around the impinging and outgoing photons' paths. Since propagating photons outward from the Alcubierre Manifold has unknown effects that may bring about anomalies, avoiding propagating 'real' photons may be critical and the need to probe the electron's state may have to be performed using virtual photons. Both approaches have to be investigated.

Both designs are shown in this paper.

The sphere depicted:

Does not require a sphere of detectors. Rather, the spheroid pattern of detection is accomplished via Near Field photons that are readily controlled using a Phased Array detection approach, which is described in a bit more detail later on. It can also be controlled via normal photons also described later via a Compton effect.

In order to 'paint' the Alcubierre Manifold first we need to extend a 'bubble' or plasma of electron-positron pairs speeding off in opposite directions. This is easily achieved at low energy by bombarding a convenient nucleus with gamma rays in the form:

$$\gamma \rightarrow e^- + e^+$$

The resulting plasma is then suspended on a static magnetic field as a plasma. Simple examples of plasmas that are not artificially contained include Earth's plasmasphere, which is enclosed by the magnetosphere:

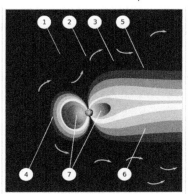

An artist's rendering of the structure of a magnetosphere: 1) Bow shock. 2) Magnetosheath. 3) Magnetopause. 4) Magnetosphere. 5) Northern tail lobe. 6) Southern tail lobe. 7) Plasmasphere.

You will note for one that the plasmasphere lies below the magnetosphere and outside the ionosphere. All sources of the magnetosphere, ionosphere, and plasmasphere come from within the Earth's rotating iron core. Thus, we have an iron core, like a churning dynamo, the magnetic field has an outer limit. Trapped within this limit is the outer edge of the plasma, sandwiched between the edge of the outer limit of the magnetic field and the ionosphere. The ionosphere is characterized by ions forming and recombining, an incomplete version of the extreme conditions of the plasmasphere. It is best to take examples from nature to show that no exotic conditions are required.

We need that electron-positron plasma to literally 'float' on the surface of this field like a bubble. Under experimental conditions, it may be found necessary to 'float' the plasma on a gas, similar to the ionosphere model.

The electron-positron pairs are directed to their respective positions along the surface of that 'bubble' using unsophisticated technology identical to a cathode ray tube, finely separating the particles into a sphere of 'pixelated dots,' for lack of a better term.

The passage of each particle, after being directed via the CRT mechanism into a sphere is then slightly different according to its spin state after passing through a type of Stern-Gerlach mechanism, and will then possess one of two possible locations on the surface of the sphere. That is, a CRT like mechanism propels them, a Stern-Gerlach mechanism then filters them prior to being propelled toward the plasma surface.

As an arbitrary example we say position 1 is occupied by electrons with spin state +1/2 and position 2 by particles with spin state -1/2. This is not unlike an old style color television where three pixels of red, green, blue were targeted to paint a single cell of color. The number of electrons hitting each pixel of RGB in the cell determined the color, the number of sweeps per second determined intensity. That is, we are using a version of antiquated technology to produce a large scale Quantum Zeno Effect.

We can actually ignore position 2 altogether and send a photon to probe position 1 only. If the photon strikes a particle, a Compton, or secondary photon is emitted and detected, meaning that the spin state of the particle, in this example, is spin +1/2. If no secondary photon is detected, that means the particle was spin -1/2. At this point, we instantaneously know the spin state of the positron at the opposite end of the engine. By using electron-positron pairs and ignoring spin -1/2 particles we've managed a meager fourfold increase in data. This may not seem like much of a gain, but it makes the difference between having to account for four quadrants of a sphere or one, our limitation being only computing power and speed.

Ironically, using current technology, the most massive and power hungry component in the system is the computer itself. In order to take the measurement and emit the photons, we use a 'phased array,' such as used with modern radar equipment. The phased array system is also antiquated technology going back to 1905. Eventually they found their use in military aircraft.

The passed array system allows one to 'paint' photons with extreme rapidity over a large volume of space, with the upper limit being set only by wavelength and computer speed.

An excellent depiction of a phased array for electromagnetic emission:

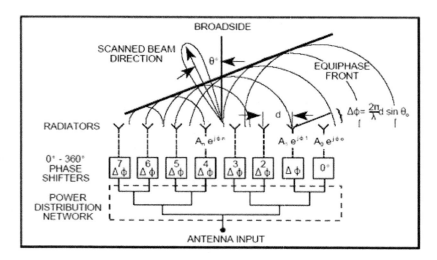

The use of the phased array means that the measurements can be taken at an extremely high rate of speed. An example of a phased array system in daily life you may be familiar with is the mechanics of the Bose speaker systems. Those tiny speakers, the way they are angled, have a time delay such that they literally create a 'virtual speaker' of large size at a given distance using almost no power. Because the 'virtual speaker' has no *real mass* the sound is far superior to the classic paper cone speaker, driven by a magnetic induction coil.

If 'real' photons are used (as in this example) the information will have to take the more mechanistic approach of having a detection system somewhere in the recoiling photon's path upon recoil. Both approaches have to be examined experimentally and the more amenable approach cannot be determined until that is done. As described below, using two phased array systems perpendicular to one another allows for the emitters and detectors to be housed together, allowing for this approach. Each functions as an emitter of outgoing, probing photons and incoming secondary photons from the particles via the opposite emitter. (see diagram below).

In the case of 'real' photons we can also use the Stern-Gerlach approach of striking the electron by a photon in its path, resulting in a Compton photon, and then again at another point of its path in a magnetic field, again resulting in a Compton photon. The shift in wavelengths of the two resulting Compton photons tell us which way the electron is curving in the magnetic field, yielding its spin state.

We can consider this one bit of *information*. This single bit of *information* is fed into our measurement rate, R, in our QZE and QAZE effect. We need to do this millions of times per second just for this one point on our plasmasphere. The number of points on our plasmasphere is dependent on the size of the vessel we need to stick inside of it, perhaps trillions, perhaps orders of magnitude more. If this is the case, we need to monitor somewhere on the orders between 10^{18} or likely more data points per second in order to paint an Alcubierre Space-time Manifold of sufficient size to circumvent a vessel. However, this limiting factor can be easily overcome by having a multitude of high speed computers each monitoring a slice of the manifold. Keep in mind, however, that 100 such computers will then have to monitor 10^{16} data points per second. The NSA monitors this amount of data routinely, who measure their data collection in petaquads (1 petaquad per second). So the technology is not unavailable or unattainable.

Reverse engineering the time coefficients determined from our first round of experiments (described a bit further on), where we measured the time coefficients with respect to the rate of observation of the quantum states of electron-positron pairs, allows us to 'paint' the Alcubierre Manifold by selecting the rate at which we take the measurements along the surface of the plasma, which is suspended like a bubble, in such a way that the given rate of measurements yields a known curvature of space.

> In addition, since we do not actually know the *resilience* of space-time under QZE conditions (if I pinch 1 mm of space-time down to 1 micron, how far does that effect extend until it is figuratively flat), we cannot assume a simple Ricci curvature will prevail under such extreme artificial curvature conditions.

It is very important to understand that to date; no hypothesis or theory has presented any means or clue whatsoever as to how the bending of space-time might be achieved by artificial means. The mechanism described here is the first hypothesis regarding the artificial bending, or 'warping' of space-time ever presented that is founded upon experimental, validated, reproducible observations in the laboratory. In addition, the power requirements to achieve the 'painting' of this space-time manifold are limited to merely taking rapid measurements, observation – only.

Oddly, the device itself becomes an anomaly in space-time because the measuring device produces a void in the space-time manifold that is unpredictable. Thus, another reason two phased arrays work opposite one another in order to 'paint' the Manifold across one another in such a way that neither is an anomaly in the space-time manifold or in a void that is not painted by measurements:

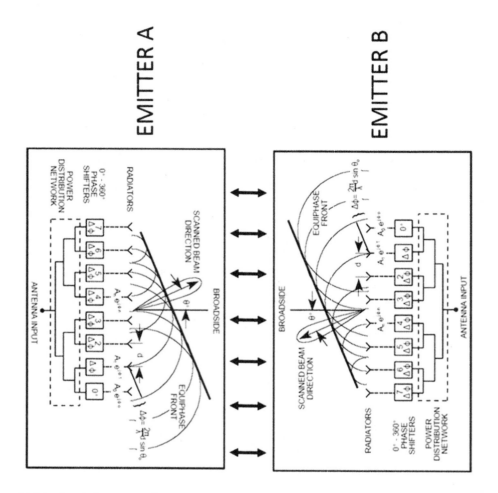

Here, emitter 'A' 'paints' the space-time manifold across emitter 'B' and vice-versa, such that neither exists in an anomalous space-time void. Emitter 'A' photons strike the electron in its path, and the Compton or secondary photon is detected at detector 'B.' In this case, having two detectors allows for the positioning of a detector in a probable location for the recoiled Compton photon as described above.

In addition, the positioning of two systems opposite one another allow both the emitters and detectors to be stationed in the two housings.

The idea of using the phased array is of course the ability to make rapid measurements by placing the photons exactly where they need to be in order to paint the manifold.

We then have to assemble all of these metrics together, the equation above for time dilation in a gravity well and the Alcubierre Metric, primarily, in order to determine the scan speed and resulting space curvature.

It is noteworthy at this point that all of the technology described up to this point is at least half a century old. The only modern construct that makes the system immediately available now is the high power and high-speed computations necessary to take such rapid measurements and control the emission and detection systems.

If one thinks of a sphere encompassing a vessel, and 'pixelating' the sphere, each pixel, denoted by an electron's path, may require millions of detection and measurement calculations per second in order to produce any noteworthy effect. In this case, we rely on the continuous climb in computing power in the industry as each year passes. As we come to the 'Coulomb' barrier of computing speed, quantum computing pics up at that point (dependent on the QZE for successful implementation) and increases our computing speed and power by orders of magnitude. The only limit to lifting mass and limit to velocity is purely computing speed and power for detection and measurement, not energy.

The first and foremost goal of this engine is lifting power. Being gravitic in nature, it defies concern for how much mass it is lifting. Greater mass requires greater computing speed for greater artificial curvature. The lifting of huge amounts of cargo, like unto that of an aircraft carrier, are for the purpose of first establishing a mining and refining infrastructure on the lunar surface for Helium-3 fusion on Earth as an inexhaustible clean energy source. The 'Faster Than Light' part of the proposal is at this time secondary, but will increase our resources to that of the cosmos, rather than being trapped on a dying planet.

THE MECHANISM THEORY AND PRINCIPLE OF THE DEVICE

The mechanics of the device involve a modified phased array for the emission of the impinging photon on one of an electron-positron pair and detection of the resulting Compton photon has any variety of remote means of detection. This is the preferred method over that of an emission of virtual photons. However, I find the positioning of two phased array systems opposite one another optimal. It is needless to say, that the challenge of producing a phased array system capable of such speed and precision will be a difficult undertaking at first, not to mention the monumental computing speed and power required, but these things are mere modifications of existing technology and certainly not beyond reach. The actual rapidity of the measurement in this prototype design will be the travel time of the resulting Compton photon to the detector and computing speed. However, as a prototype, only proof of concept is necessary. An engine producing a field of diameter 'x' will have a constantly varying Compton photon travel distance as required by the shape of the space-time manifold it produces. All of these variables have to be well characterized in order to 'paint' a manifold of the appropriate *shape*.

The electron-positron pairs are created via passing a gamma photon (a classic setup) of sufficient energy through a medium of heavy nuclei, such as the classic gold foil experiment (for illustrative argument). The electrons and positrons are then focused into a beam via the same technology as employed in a CRT television and 'sprayed' into a sphere surrounding the engine at a 'flicker rate' sufficient to differentiate the beam into individual particles such that the sphere is pixelated with electrons and positrons (at opposite ends of the engine). A magnetic field surrounds the engine such that the particles are deflected slightly according to their angular momentum (Stern-Gerlach; quantum spin state). A phased array photon strikes the particle whose position is known by timing the emission of the particle and the position of the beam (which has now crossed enough polar space to isolate the individual particles in the beam). A slight recoil of the position of the particle occurs as the impinging photon strikes it, at which point the phased array has two probable positions for the particle as predicted by its angular momentum (quantum spin state) in the presence of the magnetic field. Thus two impinging photons are emitted from the phased array and the resulting characteristic Compton wave is measured as being in one phased state (moving as depicted slightly upward and for instance, yielding a slightly shorter wavelength) or another phased state (as depicted above moving slightly downward and for instance, yielding a slightly longer wavelength). In essence, we are measuring the phase shift of the impinging photons as they recoil according to the lepton's path in a magnetic field; as a result of its spin state.

The engine will not work in any atmosphere because of the positron annihilation upon encountering any gas. The engine will only function in the vacuum of space. There is a second type of engine discussed later on, simply referred to as the type II engine, that is enclosed in a microcircuit. The type II engine is suitable for atmospheric use as it is sealed in a vacuum. Since it is enclosed in a small circuit, it curves space on a very small scale, making it useful only for heavy lifting, not for high speed.

Future engine designs will undoubtedly find any number of phenomena to measure and work around this issue.

The key to producing the Alcubierre Space-time Manifold is the balance and control of the Quantum Zeno Effect (the slowing of the progression of unitary time) and the Quantum Anti-Zeno Effect (the 'quickening' of the progression of unitary time).

By understanding these two factors, differentiating them, and characterizing them within the frame work:

$$R \propto t''$$

Where 'R' is the determining independent variable for the QZE or QAZE and t'' is the dependent variable of R.

The Quantum Anti-Zeno Effect is the speeding up of time by an alteration in observation rate. In Raizen and his colleagues work: M. C. Fischer, B. Gutiérrez-Medina, and M. G. Raizen, Department of Physics, The University of Texas at Austin, Austin, Texas 78712-1081 (February 1, 2008), if one looks at the graph presented as 'survivability' (captured ions) vs observation rate the portion of the graph in question between 5 and 30 microseconds is essentially linear: A recreation of the graphic data looks like:

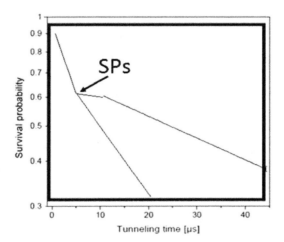

This linearity very closely supports the description above. The observation rate is a post dilation effect, not a separate function altogether, and has a linear response, as viewed head-on to the observed phenomenon. This is true of the deviation from expectation values of both the Quantum Zeno and Anti-Zeno Effects.

The point I have noted as 'SPs' is what I refer to as the 'Splitting Point,' that point in any given system where the observation rate deviates between QZE and QAZE. You will also note that the downward line is linear (no QZE at all) until it hits the SPs point. Until now, industry has focused only on the QZE with disregard for the QAZE as it has no industrial capacity as the QZE does in Quantum Computing (which uses the QZE to keep quanta from decohering (wave function collapse) before a calculation can be made). Thus, little if any reference is made to the QAZE other than being an observed phenomenon.

Using my approach to *length dilation* rather than length contraction under relativistic conditions:

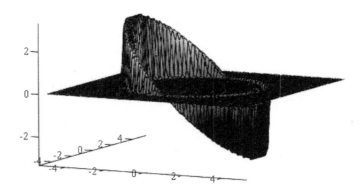

The upward spire, again, is equivalent to that we observed in the LIGOs data of two coalescing Black Holes, a space-time inversion. The downward spire represents relativistic norms. The upward spire is produced via the QZE and the downward spire via the QAZE. The flat center region, referred to mathematically as the 'tophat' function is that point just before the SPs function. The most difficult region is actually the slope running along the side, the gradient between the spires, which in turn is a gradient between QZE and QAZE functions. The mathematics are clearly described later.

Furthermore, Raizen easily achieved a t' factor of at least two, from looking at the graph. Since the speed of light is constant in all frames of reference, and is equal to one Planck unit of length divided by one Planck unit of time:

$$c = 1L_p/1t_p$$

Then

$$c = L'/t'$$

Then the observed curvature of space would then also equal two:

If

Then

$$t' = 2$$

$$c = L'/2$$

And

$$L' = 2$$

This degree of curvature far exceeds any current hypothesis on artificially producing a curvature of space by several orders of magnitude, and in fact represents an extreme velocity approaching light by an order of magnitude from an already existing validated desktop experiment. That is, Raizen's experiment clearly depicts a t' factor of two, and this *must* result in a change in Lp' by a factor of two in order to suit the demands of General Relativity. This means that the observation of a change in the progression of time can be interpreted, *must be interpreted,* as a co-occurring change in the path the information took; *a change in the shape of local space-time.* This, in simple terms, conforms to the upward spires we took a look at in the LIGO modeling data of two coalescing Black Holes that occurred for a few hundred milliseconds, a 'space-time inversion.'

In his experiment, electron tunneling was the phenomenon being measured. Electron tunneling is a phenomenon where an electron passes *through* a barrier it does not possess the energy to do so. In the arguments as I have presented them, Orthodox thinking would suggest as L' decreases, the energy requirements to cross that barrier for the electron become less. That is, at L_0, the electron does not possess the energy to penetrate the barrier. However, as L' decreases, the energy requirements become less, and the population of electrons increases. As L' increases, the energy requirements increase and the electron population decreases.

However, as shown on the graph as 'survival probability vs. tunneling time,' this is not the case. Understand that 'survival probability' refers to how many stay put and do not cross the barrier as 'survival' in the little cache where the electrons are held prior to tunneling. L' changes, but the energy requirements for electron tunneling do not. That is because there is no more *mass* of barrier to cross, just greater length, L', or less length L'. As the QZE is increased, L' decreases, and more electrons make it past the barrier. As the QAZE is increased, L' increases, and more electrons cross the barrier, *but take more time to do so.* This means that no change in energy is involved in crossing the barrier, just a greater distance is crossed, making it seem as though time were slowing, bringing us back to length dilation:

$$c = 1L_p / 1t_p$$

In General Relativity

$$l' = l_0 / \sqrt{1 - \frac{2GM}{rc^2}}$$

Special Relativity

$$l' = \frac{l_0}{\sqrt{1 - (\frac{v}{c})^2}}$$

Every quantum system (ours is the measurement of the pin states of electron positron pairs) will have a unique Zero Point based on a lower limit of Planck's constant, and involves many variables such as the system, the local space it occupies, the conditions of the space it occupies, the means of detecting what the system is doing, and so on. In order to determine the true Zero Point of any given system, one has to take all of these variables into account and refer to Quantum Electro Dynamics in order to actually perform the calculations. The calculations and work are rather extensive.

The Zero Point is not *directly* necessary to the development of the engine. It is however a start of cataloging QZE systems for future development of more powerful engines.

However, this Splitting Point doesn't directly coincide with the Zero Point of the system. It, according to Raizen's data, just upon visual inspection describes electrons tunneling across a barrier in one of two time frames. Treating it like a true graph, it is a split in linearity. It is behaving so much like a Zero Point affected by a *hidden variable* and it should be treated as such. It is best derived experimentally, typically, since we do not have a model for any QED calculation for any QZE system to date. That is why I suggest beginning mapping out such data.

Therefore, I will simply generalize the above example by referring to the Zero Point for this system that involves electrons tunneling across a barrier under the conditions of the experimental design as:

ZP$_s$

This will be our generic Zero Point for the system in a *purely temporal mechanics approach* but will have to resort to an alternate description, with the 's' subscript referring to the 'split' between the two legs of the graph toward QZE and QAZE, and refer to it as 'SPs.'

This term refers to the point of the rate of observation, which will be designated 'R,' where the Quantum Zeno Effect and Quantum Anti-Zeno Effect are observed to deviate from 'normal' time as depicted in the graph above, before the graph splits. A 'split' occurs where the observed phenomenon behaves in a fashion that displays deviating from the 'normal' expected observation of the progression of time as designated for the system. For radioactive decay, quantum tunneling, etc., that point where the normal progression of time as defined for the system deviates from the defined path and appears to have a different rate of progression with respect to time; and may be observed to progress faster or slower, centered on that exact observation rate, 'R' at 'SPs'.

These values have no current mathematical description and have to be determined for each type of system via experimentation. That is, there is at this time no theoretical model to predict such values and all such variables and constants will have to be experimentally derived for each type of system employed. Again, I expect differing systems, such as radioactive decay and electron tunneling to have remarkably different variables and constants. However, I expect two different forms of beta decay, for instance, to have similar but not identical values. The fact that tritium and cesium have different half-lives is an indication of temporal differences at the grass roots level. The relationships between observation and probability have been an awkward and noticeable subject for over a century.

The relationship between what I am referring to as the Splitting Point, named because of the apparent and obvious split in Raizen's graph of tunneling electrons, and Zero Point, is a fundamental shift in ground state of energy. That is, Raizen's experiment, using tunneling electrons, showed this shift in ground state most obviously and clearly. The graph in question, shown here again:

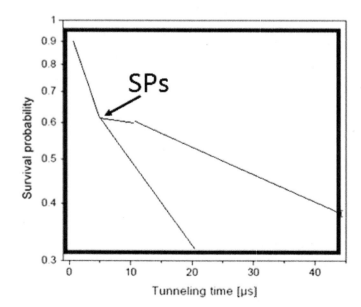

- Fischer, M.; Gutiérrez-Medina, B.; Raizen, M. (2001). "Observation of the Quantum Zeno and Anti-Zeno Effects in an Unstable System". Physical Review Letters 87 (4): 040402.]

- Raizen, M. G.; Wilkinson, S. R.; Bharucha, C. F.; Fischer, M. C.; Madison, K. W.; Morrow, P. R.; Niu, Q.; Sundaram, B. (1997). "Experimental evidence for non-exponential decay in quantum tunnelling" (PDF). Nature 387 (6633): 575.

Demonstrates electrons tunneling over a barrier that they lack the energy to tunnel through or otherwise failing to tunnel through the barrier as a result of observation rate. This represents a fundamental shift in basal energy of the individual and collective electrons. Shorter observation times will approach the Heisenberg barrier, where delta-E will become significant:

$$\Delta E = \frac{h}{2\pi \Delta t}$$

Which becomes a remarkable result, because here, delta-t is for the observer, not the electron. However, *the electron knows it is being observed.* This phenomenon has led to a century of debate and resulted in the controversial Von Neumann Copenhagen Interpretation of Quantum Mechanics.

In a double-slit experiment you shoot a single electron at the double slits. Even though there's only one electron, it still forms an interference pattern, as they build up over time. It's as if the electron travels through both slits simultaneously. This demonstrates the electron as a wave function.

However, just by *observing* the double-slit experiment, the behavior of the electrons changes. A century of work has gone into finding ways to passively observe the experiment without affecting the experiment, so the outcome is confident that the detection is not interfering with the experiment. *Observation* of the experiment *causes* the interference pattern to fail to appear.

The idea behind the double-slit experiment is that even if the electrons are sent through the slits one at a time, there's still a wave present to produce the interference pattern. The wave is a wave of probability, because the experiment is set up so that the scientists don't know which of the two slits any individual electron will pass through.

However, deeper still is that when we pass one or more 'particles' and get a wave function result from an electron or electrons, we have taken a spin ½ particle and in order to see an interference pattern that spin must become either zero or 1, there is no work around for this. Of all the volumes of marvels I have read about the double-slit experiment, I have never heard this mentioned, perhaps it is too obvious to point out. We know nothing about what spin is, other than that it is the most exploited characteristic in entanglement arguments.

But if they try to find out by setting up detectors in front of each slit to determine which slit the electron really goes through, the interference pattern doesn't show up at all. This is true even if they try setting up the detectors behind the slits. No matter what the scientists do, if they try anything to observe the electrons, the interference pattern fails to emerge. *The spin remains ½?* But this is true for photons as well. Nonetheless, turn your eye toward the experiment and that electron spin becomes either zero or 1.

This, and not all of the banter regarding particle vs. waves, is the mystery of the double slit experiment, and the argument that drove the founders of Quantum Theory, particularly Von Neumann, to the conclusion that the *observer* is *interdependent* with the system. This is referred to as the Von Neumann Copenhagen Interpretation of Quantum Mechanics.

Unfortunately, it required, as many of the arguments in deep Quantum Theory did at the time, a hard definition for *consciousness* suitable within the framework of Quantum Physics. This is why I supplied such in this paper.

That is, in the Heisenberg equation above, the Δt is the observation rate of the observer, and the ΔE is the energy borrowed by the electron as a result of observation, in Raizen's experiment. We can also apply this same principle to the QZE and QAZE in general. In this case, I am referring to equating the QZE and QAZE to Special and General Relativity.

Thus, this Splitting point between the Quantum Zeno Effect and the Quantum Anti-Zeno Effect may have consequences on a system with respect to its Zero Point if delta-t becomes small enough to affect significant change in delta-E.

Since the system has a Splitting Point based on Planck's constant (based on the same principle as the Zero Point, but as of yet lacking a mathematical description on this scale), then there will also be a Splitting Point for this system with respect to the Planck unit of Length, Lp, and the Planck unit of time, t_p.

In the graphs above it is apparent that there is a consistency in the activity of the system and then a sudden departure from the 'normal' expected behavior of the system splitting off into two branches. The point at which this splitting of the two branches occurs will then be the respective Splitting points for the Planck unit of length, Lp, which I will refer to as SP_{Lp} and a Splitting Point for the Planck unit of time, t_p, which I will refer to as SP_{tp}.

This exact principle will be used to define the shape of space-time in order to produce the Alcubierre manifold, which requires both a shortening of space-time and a lengthening of space-time to manifest.

It is commonly thought that the QZE is a purely temporal phenomenon. That is, that the progression of time is being altered as a result of the progression of time chosen to measure or observe the phenomenon. The temporal aspect of the QZE is already accounted for in the rate of observation (the so-called measurement problem, of which there are countless hypotheses). The secondary effect cannot be another temporal effect but must be a deviation in the path of the information to the detector; a *spatial effect as required by General Relativity and Special Relativity.*

To imply a secondary effect on the rate of the progression of unitary time within the observed system as a result of the rate of the chosen progression of observed time of the system (observation rate) by the observer is why the phenomenon is not understood. That is, of all of the hypotheses I have read concerning the measurement problem, I have not seen one that describes *why* the observation rate and the apparent progression of unitary time differ. In all cases the best mathematical arguments come up with equivalence.

It is a re-iterative argument. There is a realm of confusion regarding the QZE because of this re-iteration of the rate of observation, 'R,' and the rate of progression of time for the system. This is the result of reading about the phenomenon but not understanding what is written. Therefore, consider the following:

> If one observes a radioactive cesium mass, cesium being the primary means by which we measure time, at 1 billion observations per second, bordering on *constant observation,* the mass of cesium *will stop decaying altogether, indefinitely. The mass of cesium atoms that we use to measure time has stopped decaying, telling us that time has **stopped** for the cesium mass's local environment.*

As we slow down the observation rate the cesium mass begins to decay again, slowly at first, meaning that time is still running slow, but is gradually picking up speed as we slow the observation rate. At a certain point, we reach SP_s, the Splitting Point in the observation rate, and the cesium mass begins to decay *faster than normal,* meaning that time as measured by cesium decay has 'sped up' in the cesium mass's local environment. Cesium is used as the absolute measure of time by the Bauru of Standards and NIST.

There is no issue of any 'measurement problem' here. This experiment has been done. The radioactive decay of cesium was used to confirm Special Relativity, and later (when rocketry became available) to confirm General Relativity to high accuracy.

At this time, I am not willing to commit to the following statement, but in a sense the QZE, that is, the observed effect on the progression of time, and the causal relationship because of the rate of observation, seem to take on a self-similar relationship. Where 'R' is the rate of observation, and QZE is the rate of progression of time because of 'R' as a fractal characteristic:

$$R \underset{\alpha}{\leftrightarrow} \text{QZE}$$

That is, the causal relationship *may not be* entirely unidirectional, at least for a local environment. We, at this time, have no idea how the QZE affects non-local environments. For instance, in General Relativity gravity extends for infinity, albeit we can say the effects are negligible at great distance, our distance to our engine will not be 'great distance.'

As the observation rate, for instance, slows the observed rate of the progression of time for the system, since we do not know how far reaching this effect is, may affect the observation rate for the observer at least out to some distance, in turn affecting the progression of time for the observed system, and so on. This would lead to an indiscreet value for the QZE, and as demonstrated in a Mandelbrot set, could lead to wildly deviating values from linearity. Since the selection in the system regarding who is the observer and whom is the observed seems to be arbitrarily associated with the higher mammal, formalizing Von Neumann's Subjective Wave Function Collapse Theory, AKA Copenhagen Interpretation of Quantum Mechanics, there is no other formalism in Quantum Mechanics that assigns either system as the observer or the observed. Therefore, the assignment of the higher mammal as *causing wave function collapse is true, and/or:*

$$R \underset{\alpha}{\leftrightarrow} \text{QZE}$$

Is a true statement that leaves the observed/observed system as arbitrary. As evidence for this, I present the first actual real time graph of data collected by Raizen and his group of tunneling electrons; over a mere 40 microsecond observation range there is both a split in time deviating from t'>1 to t'<1 and a kink in the t'>1 leg of the graph.

That is, the selection of the observer is arbitrary, our detector for that system designated as 'R.' This designation is formalized in Quantum Mechanics only by the Von Neumann Subjective Collapse Theory, the Copenhagen Interpretation of Quantum Mechanics. However, the higher mammal also designates the system to be observed, the QZE system. As the detection rate for R increases, the progression of linear time for the system, QZE appears to slow. However, what does the system, QZE observe of the system detector, R? If we equate the phenomena with Special and General Relativity, the QZE system should perceive the system detector R as *quickening*. This cycle of mutual perception can go on indefinitely. Later in the text Quantum Temporal Dynamics, I will describe the fractalization of time as a phenomenon of Gravitation, which follows the rule:

$$t' \to \infty$$

And/or

$$t' \to 0$$

In the case of gravitation, using the extremes of a Black Hole as an example, one observer, t' approaches infinity, is nearing the Swarzschild Radius, the other observer is at a distance where t' approaches flat space or zero. Without both values and both results, the cosmos cannot exist; space-time would not be curved here but quantized to two huge values.

Both systems, R and QZE, regardless of being arbitrarily selected as cause and detector (independent and dependent variable) by the higher mammal; nature, even if obeying the Von Neumann Interpretation completely, still does not regard the *whim* of the higher mammal, and sees the selection as arbitrary. Thus, the causal relationship in the QZE experiment becomes a bit cloudy, and the linearity of the dynamics of the system can be expected to take on a pattern of self-similarity as:

$$R \underset{\alpha}{\leftrightarrow} QZE$$

Therefore, navigating the linearity of such relationships between R and the QZE can and will become difficult.

The proper argument is that the rate of observation is the temporal effect, and the resulting *observed* effect is that the local shape of space-time is altered, causing the information regarding the state of the system to take a different (longer or shorter) path to the observer (detector). *However,* this description is capable of fitting both R and QZE independently and simultaneously.

In our example of Cesium decay, it is just as easy to say that as the observation rate, R is increased, the distance to the detector becomes a longer path.

The QZE is the effect of the shape of space as a result of the choice of the chosen progression of time of the measurement, as is demanded by the rules of General Relativity and Special Relativity. According to my argument above, that as time dilates space dilates, and (for lack of a better term) as time anti-dilates space contracts, the QZE is a sensible reaction of space to the temporal aspect of the chosen progression of time of the measurement.

In other words, in line with my remarks suggesting that the demanding requirement $c=1L_p/1t_p$, as time is *apparently dilated, the cause is that t_p' has contracted,* the value L_p must contract. In this case, information, such as the passage of a decay particle from a radioactive source, has to cross a greater number of Planck units of length, because each Planck unit of Length is now smaller, making the apparent path of the decay particle having to cross a greater distance, as measured in Planck lengths, making it appear as though it is 'slowing down,' when in fact it is taking a *longer path to the detector in terms of Planck intervals of space it has to cross.* The mathematical description for this is described below.

There is a matter of semantics here that needs to be cleared up in the use of Lorentz upside-down equations. If I extend the Planck length 4.4 light-years to the nearest star, Alpha Centauri, in one sense we say that we have increased Lp' to 4.4 light-years. However, to an observer slightly off center to one side, he/she observes Alpha Centauri one 'normal' Planck length distance, i.e. closer, *contracted.* That is the *observed* effect. The real effect is extending one Lp to 4.4 light-years. If you take the twin paradox and turn all of the Lorentzian arguments upside-down, it still works. For the traveler, the distance to Alpha Centauri is one Planck length. As for what a third party sees, he/she, like the Bob/Alice EPR paradox sees you Superpositioned at both points simultaneously.

As for the Quantum Anti-Zeno Effect, our chosen reference for the progression of time of observation is *longer, dilated,* resulting in length dilation. In this case the Planck unit of length is longer and the decay particle has to cross a lesser number of Planck units of lengths, *taking a shorter path to the detector,* making it appear as though the rate of decay is accelerated. (This Quantum Anti-Zeno Effect, as will be mathematically described as a lengthening of the value Lp, is the primary principle of the engine's design, to increase the value of Lp, *stretching the nose of the ship to the target*).

If we take a closer look at the definition for t_p':

$$tp' = \sqrt{\frac{hG'}{2\pi c^5}}$$

And the definition for G':

G' = 6.67384(80)×10^{-11} m^3/Kg (t')^2

Here it is correct to state that as t' increases G' decreases, and as t' decreases G' increases.

As we substitute t' for 'time dilation:

$$f(tp') \lim_{t' \to \infty} \sqrt{\frac{hG'}{2\pi c^5}} = 0$$

And it can be summarized that:

$$as\ t' \to \infty;\ G' \to 0;\ t_p \to 0$$

Meaning that as time dilation ($t' \to \infty;\ G' \to 0;\ and\ t_p \to 0$) is the result of t_p diminishing; *time dilation is the result of information having to cross a greater number of Planck intervals of time to reach its destination.*

The same exact principle holds true then for the Planck length, Lp: given G' = 6.67384(80)×10⁻¹¹ m³/Kg (t')²

$$f(Lp') \lim_{t' \to \infty} \sqrt{\frac{hG'}{2\pi c^3}} = 0$$

And it can be summarized that:

$$as\ t' \to \infty;\ G' \to 0;\ Lp \to 0$$

The effect of length dilation is the result of information having to cross a greater number of Planck lengths in order to reach its destination.

And all of the reciprocal values hold true; given G' = 6.67384(80)×10⁻¹¹ m³/Kg (t')²

$$f(tp') \lim_{t' \to 0} \sqrt{\frac{hG'}{2\pi c^5}} = \infty$$

And can be summarized that as:

$$as\ t' \to 0;\ G' \to \infty;\ t_p \to \infty$$

And for the Planck length:

$$f(Lp')\lim_{t'\to 0}\sqrt{\frac{hG'}{2\pi c^3}} = \infty$$

And can be summarized as:

$$as\ t' \to 0;\ G' \to \infty;\ Lp \to \infty$$

In order to understand the statement, $t' \to 0$, we look at the reciprocal values for t', since the Lorentzian equation has been upside down for 100 years as *an observed effect:*

$$tp' = \frac{t_{p0}}{\sqrt{1-\left(\frac{v}{c}\right)^2}}$$

The *real* value of t_p has changed (diminished), resulting in the observed effect:

$$tp' = tp_0\sqrt{1-\left(\frac{v}{c}\right)^2}$$

The reciprocal values differentiate the observer and the observed, eliminating any 'paradoxes.'

And it can be summarized that:

$$As\ tp' = \frac{t_{p0}}{\sqrt{1-\left(\frac{v}{c}\right)^2}} : \geq 1\ and\ t' \in \{1...\infty\}$$

$$and\ as\ tp' = tp_0\sqrt{1-\left(\frac{v}{c}\right)^2}\ t' \leq 1\ and\ t' \in \{0...1\}$$

The first case represents the stationary observer observing the speeding traveler. The second case represents the speeding traveler observing the stationary observer. There are no *paradoxes* when the equations are presented in this way. The first case represents the summarized form (The first case represents the stationary observer observing the speeding traveler, actually observing a Superpositioned traveller):

$$as\ t' \to \infty;\ G' \to 0;\ t_p \to 0$$

Where *the real value t_p has diminished, requiring that the information cross a greater number of Planck units of time and Planck lengths to reach its destination; according to:*

$$as\ t' \to \infty;\ G' \to 0;\ Lp \to 0$$

The information has to cross a greater number of Planck lengths in order to reach its destination.

The second case, the second case represents the speeding traveler observing the stationary observer:

$$as\ t' \to 0;\ G' \to \infty;\ t_p \to \infty$$

The information crosses a lesser number of Planck intervals of time to reach its destination according to:

$$as\ t' \to 0;\ G' \to \infty;\ t_p \to \infty$$

And

$$as\ t' \to 0;\ G' \to \infty;\ Lp \to \infty$$

That is, in the extreme case where v=c, or very nearly so, both Lp' and tp' have increased to some indefinite length and the information only has to cross one Planck interval of space-time to reach its destination, never exceeding the hard definition for the speed of light, c=1Lp/1tp, and also following our previously described definition for quantized motion. Again, the third party, Bob/Alice observes a superpositioned traveler in two apparent locations simultaneously.

This increase in length follows the same rules as our definition for quantized red-shift:

$$f_{observed} = f_{emitted} \sqrt{\frac{1-((nLp/xtp)/(1Lp/1tp))}{1+((nLp/xtp)/(1Lp/1tp))}}$$

At v=c, *f=0*, or, Superpositioned. That is, at v=c, all time and distance in the cosmos is zero, and the photon is said to be Superpositioned throughout space-time. It cannot have a frequency, designating any proper time. Thus, *f=0*. This represents *constant observation* of that wave function, regardless of preferential perspective. Her go, the century old equation for relativistic redshift describes the QZE. In this form, I have merely quantized it.

As for: (considering the correction of length dilation rather than length contraction) *using the corrected form*

$$Lp' = Lp / \sqrt{1 - \frac{2GM}{rc^2}}$$

And time dilation in a gravity well:

$$t' = t_0 / \sqrt{1 - \frac{2GM}{rc^2}}$$

$$as\ t' \to \infty;\ G' \to 0;\ t_p \to 0$$

$$And$$

$$as\ t' \to \infty;\ G' \to 0;\ Lp \to 0$$

Here, if we take into consideration that 'r' (the distance from the center of mass) as measured either in Planck lengths or Planck units of time (by the transformation c=1Lp/1tp, the ration G'/r remains fixed. In this case the statements:

$$Lp' = Lp / \sqrt{1 - \frac{2GM}{rc^2}}$$

And

$$t' = t_0 / \sqrt{1 - \frac{2GM}{rc^2}}$$

Hold true, if and only if we take Lp' as dilating rather than contracting.

It is important and confusing to hold all of these definitions up in the air in one's head, referring to dilation and contraction being 'upside-down,' in one sense, and backward. Therefore, we keep the generalized statements in mind:

$$as\ t' \to \infty;\ G' \to 0;\ t_p \to 0$$

$$And$$

$$as\ t' \to \infty;\ G' \to 0;\ Lp \to 0$$

Time dilation is the effect of tp' diminishing and Lp' diminishing, resulting in information having to cross a greater number of Planck units of time and length to reach its destination; the result is an apparent slowing of time as the information takes a longer path to the observer.

The same holds true for gravitational red-shift, given by:

$$Z(r) = \frac{1}{\sqrt{1 - \frac{2G'M}{c^2 r}}} - 1$$

Given that:

$$as\ t' \to \infty;\ G' \to 0;\ Lp \to 0$$

The ratio G'/Lp remains fixed resulting in G'/r remaining fixed; 'r' is now a fluid value, a function of Lp'. The value 'r' will be fluid and have a set of quantized values that change with the apparent distance from the source or center of the gravitational phenomenon.

Going back to the differentiation point between the Quantum Zeno Effect and the Quantum Anti-Zeno effect, the exact point where the path of the decay particle, tunneling phenomenon, and so on, departs from common time is the 'Splitting Point' for the system (again, similar to the Zero Point but as of yet lacking a sufficient mathematical description that must include the observer and method of observation as well as rate of observation), resulting in an alteration of SP_{Lp} and SP_{tp}.

With respect to the 'Splitting' Point for the system, we define the SP_{tp} as:

$$tp' = \frac{t_{p0}}{\sqrt{1-\left(\frac{v}{c}\right)^2}} \text{ if } SP_{tp} > 1$$

And

$$tp' = tp_0 \sqrt{1-\left(\frac{v}{c}\right)^2} \text{ if } SP_{tp} < 1$$

Both states can be inverted as representing some preferential perspective of the system being observed or the 'real' state of the local system. Lp' can be treated the same way, of course.

This is the point where the two possible paths of information, one path crossing a greater number of Planck lengths to the detector as a result of Lp' becoming 'smaller,' resulting in apparent slowing of time, the other taking a shorter number of Planck lengths to the detector as Lp' increases, resulting in an apparent quickening of the progression of time.

THE ENGINE DESIGN

In the following proposed engine design, the quantum spin states of electron-positron pairs are measured at very rapid rates via photon excitation in a magnetic field and subsequent detection of the phase shift of the resulting Compton wavelengths of those photons as they recoil from the leptons that are moving along slightly different paths according to their angular momentum. This is essentially a Stern-Gerlach setup that replaces a stationary detector with the measurement of the slight phase shift of Compton photons from leptons taking slightly different paths in a magnetic field.

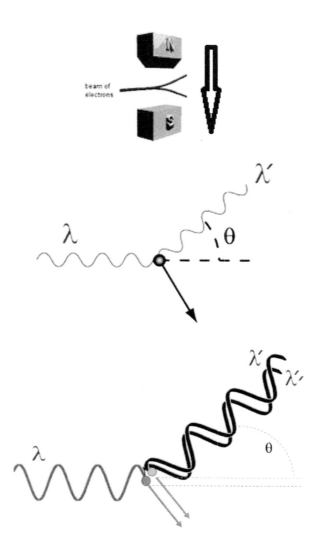

The difference in the paths of the spin +1/2 and spin -1/2 particles will be detected by the phase shift in lambda' and lambda''.

The electron spin state as being either +1/2 or -1/2 is chosen as the measurement because it can readily be concluded that the quantum spin state of a lepton cannot be altered as a result of the rapidity of the measurement. Electron-positron pairs are selected because the entangled feature of passing them in opposite directions away from the engine source allows the measurement to be focused on one end of the field while simultaneously gathering information regarding the opposite end of the field around the engine source. Also, manipulation and the mechanism of control of electron positron pairs is century old technology.

It is also important to note that only two measurements have to be taken for any given electron in order to determine its path in the magnetic field. There is no precedence for considering a larger number of measurements of the same electron are necessary to produce a true Quantum Zeno Effect. Furthermore, it would seem conclusive that in Raizen's QZE experiments for any given sodium ion that tunneled over the energy barrier only a measurement of the ion's condition before it tunneled and its condition after it tunneled were taken (two measurements), or otherwise necessary; not a sustained observation of any given ion prior to or during quantum tunneling and during the process of being trapped.

The resulting Quantum Zeno and Anti-Zeno effect will be measured by determining the resulting curvature of space around the phenomenon via measuring the deviation of the Compton red-shift photons from the predicted value according to the demands of General Relativity, and this is the working principle of the engine.

The first phase of the experiment will be to characterize the resulting curvature of space in order to 'paint' a space-time manifold similar or identical to that which was proposed by Alcubierre:

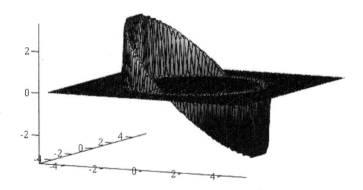

Both the Quantum Zeno and Anti-Zeno Effects that information reaching the detector in two paths deviating from the predicted, one shorter path and one longer path, as marked by the acceleration or deceleration of the progression of time, *demand* that the shape of local space is altered according to the laws of General Relativity.

In this engine design, because of the redefining of tp' and Lp' the Alcubierre space-time manifold as I present will appear and be described in terms that are both upside-down and backward from the classic depiction of this manifold as a result of the correction to the Lorentz equation being inverted.

First, it is necessary to determine the time coefficients for the phenomenon, in this case, of observing electron-positron spin state pairs. Again, once the measurement is taken for the quantum spin state of the electron, for instance, the quantum spin state of the positron is known instantaneously, and will always be opposite that of the entangled antipode.

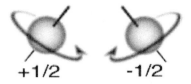

For the meantime, we will designate a coefficient for the alteration of the progression of unitary time via the observation of quantum spin states for entangled electron-positron pairs as 'k.' (Please do not confuse this with the Kaon).

We will agree on a designation of $+k$ as representing a slowing of unitary time by taking more rapid measurements (sustained observation, Quantum Zeno Effect) and $-k$ as a 'speeding' up of unitary time as a result of taking less rapid measurements (release, or Quantum Anti-Zeno Effect) based on the experimental data we have first acquired regarding the system's 'Splitting Point' between the Quantum Zeno and Quantum Anti-Zeno Effects, which would occur at k^0.

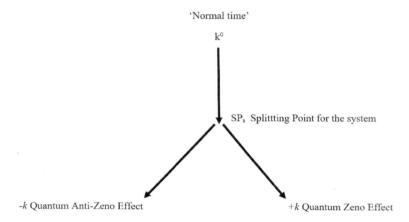

The value, k^0 remains in space 'normal time' until the observation rate; 'R' reaches the critical value, R_{sps}, where a deviation in the observed rate of the progression of linear time for the system is observed.

For the sake of argument, we will agree for the duration of the argument that altering the rate of progression of unitary time alters the shape of space. It is not necessary that the shape of space is altered in the same way or to the same degree as for *time dilation*, only that it is altered. That is the reason for the re-defining of the Einstein-Ricci space-time curvature as:

$$\frac{8\pi G'}{Lp^4/tp^4} T'_{ab}$$

Here, G' has been described at length, T' has to be determined experimentally and is non-predictable and likely unique to each system.

In this or any other mathematical scenario, it does not matter what the relationship is, only that it becomes mapable, and as such predictable. As such, the alteration of the observation rate, 'R' will bring about a predictable alteration in the shape of highly localized space, as is our goal in producing a space-time manifold.

In this re-definition of the Einstein-Ricci space-time curvature it is noted that the term L_p/t_p unambiguously quantizes space-time, thus quantizes G', and quantizes T', thus quantizing the space-time curvature.
In order to 'paint' the Alcubierre space-time manifold as depicted:

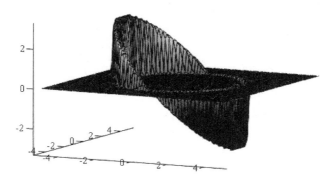

The requirement will be to take measurements across the entire domain of the manifold of entangled electron-positron pairs. In this depiction, the lower curve represents space-time dilation, slowing time, values of $+k$, by taking more rapid measurements (sustained observation). The upward curve represents space-time contraction, increasing the rate of the flow of unitary time, designated $-k$ (by taking less rapid measurements (release, Quantum Anti-Zeno Effect). *The inward or outward curvature of space is dependent on the corrected argument for length dilation vs length contraction, and merely results in the orientation of the emitter to be turned 180 degrees.*

In order to quantize Alcubierre's manifold to match our re-definition of the Einstein-Ricci space-time curvature, we'll make some minor adjustments to Alcubierre's original form of the equation such that:

$$-\frac{c^4}{8\pi G}\frac{v_s^2(y^2+z^2)}{4g^2 r_s^2}\left(\frac{df}{dr_s}\right)^{'2}$$

The term c^4 becomes (L_p^4/t_p^4)
The term G becomes G'
The term v_s^2 can take the form $((nL_p/xt_p)/(1L_p/1t_p))^2$
The term $(y^2 + z^2)$ becomes $(L_{py}^2 + L_{pz}^2)$
The determinant metric g^2 will be treated like T' and simply referred to as g'^2
The term r_s (radius of the center of the phenomenon) is represented by $r_s(t) = [(x-x_s(t))^2 + y^2 + z^2]^{1/2}$ essentially replaced by quantized values such that

$$[(L_px - L_px_s(t'))^2 + L_{py}^2 + L_{pz}^2]^{1/2}$$

And since our term for *f* is represented by

$$f(r_s) = \frac{\tanh(\sigma(r_s + R)) - \tanh(\sigma(r_s - R))}{2\tanh(\sigma R)}$$

Here, 'R' is not to be confused with the detection rate, 'R' that we discussed earlier, but references a 'top hat' function, leaving us with that nice flat domain in the center of the manifold. The term *sigma* refers to the wall thickness of the warp bubble and 'R' as the total radius of the warp field.

In addition, provided we remain consistent with the former two definitions of r_s and f (since sigma and 'R' are arbitrary values by definition) we can rewrite, or rather substitute the values in such that:

$$-\frac{(Lp^4/t_p^4)}{8\pi G'} \frac{((nLp/xt_p)/(1Lp/1t_p))^2 (Lp_y^2 + Lp_z^2)}{4g'^2[(Lp_x - Lp_{x_s}(t'))^2 + Lp_y^2 + Lp_z^2]} \left(\frac{df}{dr}\right)^2$$

We then look at two relationships:

$$R\alpha \pm k$$

And

$$R\alpha t''$$

Moreover, I believe at this point, taking our coefficient +/-k into account, we can say that the Alcubierre manifold as depicted in this form 'maps' the element a (+/-k) into the element b (the very large equation above) such that:

$$f(\pm k) \mapsto -\frac{(Lp^4/tp^4)}{8\pi G'} \frac{\left(\frac{nLp}{xtp} \atop \frac{1Lp}{1tp}\right)^2 (Lp_y^2 + Lp_z^2)}{4g'^2[(Lp_x - Lp_{xs}(t'))^2 + Lp_y^2 + Lp_z^2]} \left(\frac{df}{dr}\right)^2$$

This form of the equation merely states that the function 'f' maps the element +/-k into the element 'b,' which in this case is the quantized form of the Alcubierre metric, representing a population of the set of particles being used to describe the manifold.

However, as I've stated before, in this space-time manifold, the Lorentz equation is originally inverted, thus we drop the negative sign:

$$f(\pm k) \mapsto \frac{(Lp^4/tp^4)}{8\pi G'} \frac{\left(\frac{\frac{nLp}{xtp}}{\frac{1Lp}{1tp}}\right)^2 (Lp_y^2 + Lp_z^2)}{4g'^2[(Lp_x - Lp_{xs}(t'))^2 + Lp_y^2 + Lp_z^2]} \left(\frac{df}{dr}\right)^2$$

The function 'k' represents the rapidity of the measurement (Rα±k), the angle along the plane of measurement, *and the distance to the measured phenomenon, keeping in mind that measuring any phenomenon at a distance greater than 1Lp is an observation of an event that occurred in the past, not a 'co-moving' present.*

I say comoving because if the working principle of the engine is in fact a superpositioning phenomenon, then comoving is the proper terminology to describe apparent motion via the manifold.

The biggest problem with the above equation is the value 'pi.' It cannot be quantized regardless of the approach. Since the construction of the manifold may rely heavily on the production and annihilation of virtual photons, pi will be replaced with '3.' (This is true of near field physics)

$$f(+/-)k \mapsto \frac{(\frac{Lp^4}{t_p^4})}{24G'} \frac{\left(\frac{\frac{nLp}{xt_p}}{\frac{1Lp}{1t_p}}\right)^2 (Lp_y^2 + Lp_z^2)}{4g'^2[(Lp_x - Lp_{xs}(t'))^2 + Lp_y^2 + Lp_z^2]} \left(\frac{df}{dr}\right)^2$$

In the case where 'real' photons are used the value 'pi' simply becomes a problem for quantization on a quantum scale and has to be approximated to the best experimental degree.

THE MEASUREMENT RATE 'R'

Thus we have the rate of measurement, 'R', the distance to the observed particle, giving it a unique name L_{op}, which defines our phenomenon of measuring an event that occurred *in the past,* and a unique angle of the measurement in a sphere around the engine, 'S.' These values make the function '$\pm k$', which can take on a positive or negative number, in *my depiction* of the manifold, k is positive when the measurement rate is sufficient to slow the progression of time, and negative when it is sufficient to speed the progression of time.

$$for + k; \ as \ t' \to \infty; G' \to 0; t_p \to 0$$

$$And$$

$$for + k; \ as \ t' \to \infty; G' \to 0; Lp \to 0$$

And

$$for - k; \ as \ t' \to 0; G' \to \infty; t_p \to \infty$$

$$And$$

$$for - k; \ as \ t' \to 0; G' \to \infty; Lp \to \infty$$

The mainstream approach is thinking in terms of contracting space in front of the engine. However, as I've pointed out that would lead to $Lp \to 0$ and we have to travel an infinite distance (in terms of number of Planck lengths) to gain no forward motion. The correct approach is to extend the value Lp out to some indefinite distance ($Lp \to \infty$) such that the 'nose' of the engine approaches the destination (this is where the term comoving comes in), and snap the rear of the engine back to the destination, the rear defined as $Lp \to 0$ then collapsing the function such that $Lp \to 1$.

The next step is to define our time coefficient, +/- k into the measuring of the space inside the manifold by taking measurements in the entire volume of space. However, our term *volume* has taken on a new meaning; the *volume* is now defined by our time coefficient +/-k, and not subject to our former vision of the given volume of space in question. The actual *volume* will be defined by taking measurements on the outer surface of this manifold (like the surface of a bubble or balloon) at rates defined by +/-k in order to define a *new volume, and a new shape for space-time within the manifold.*

However, to the detector, the manifold will always appear as a sphere. Using the double phased array approach, which is the ideal arrangement, the plasma will always appear as a sphere, however, the space-time manifold will be structured as described by Alcubierre's equation, albeit with some corrections.

In oversimplified terms, if we look at the metric from the center and make a radial measurement towards the forward leading edge, depicted by the arrow (again, the portrayal is inverted from its typical depiction):

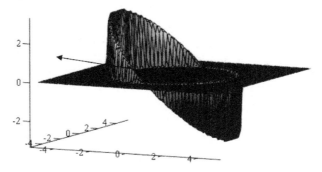

The measurement, starting at the center of the manifold, originating from the origin of our 'line-of-sight' as depicted by the arrow, will change the rate of measurement which we will designate R_m according to the time coefficient, in this case varying $+/-k$ with the rate of measurement occurring in the regions; *for one measurement of one quantum spin state for one particle, in one dimension.*

Again, we are using this phased array system by emitting photons:

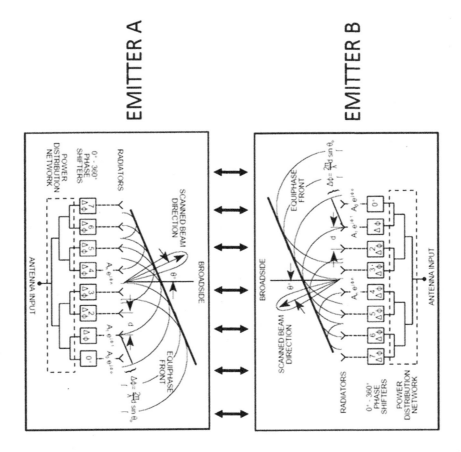

To look at these quantum spin states

Which have been finely divided and formed onto a plasma around our engine core:

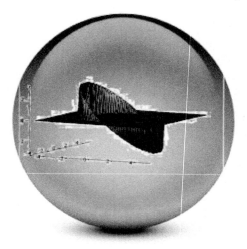

The quantum spin states of the leptons have been separated by a modified Stern-Gerlach type setup suitable for the process:

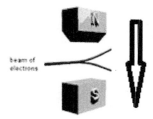

The phase shift of the Compton recoil photons is the data collected:

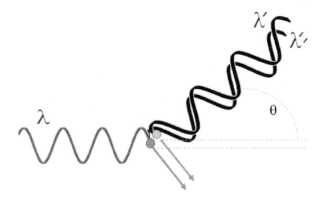

The processing speed at **Rα±k** according to

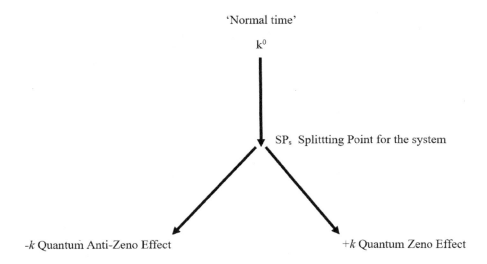

$$f(+/-)k \mapsto \frac{(\frac{Lp^4}{t_p^4})}{24G'} \frac{\left(\frac{\frac{nLp}{xt_p}}{\frac{1Lp}{1t_p}}\right)^2 (Lp_y^2 + Lp_z^2)}{4g'^2[(Lp_x - Lp_{xs}(t'))^2 + Lp_y^2 + Lp_z^2]} \left(\frac{df}{dr}\right)^2$$

Defines the shape of the Alcubierre Space-time Manifold:

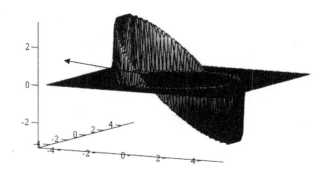

The property of the manifold which has been overlooked that prevents numerous problems is that the manifold is painted both above and below the 'ship,' encapsulating it, like air passing above and below a wing such that space bends around the vessel:

Our capsule, of course, does not possess aerodynamic properties as depicted in this 'cut and paste,' but is symmetric, and is certainly not intended to perform any lift. This 'cut and paste' image is merely intended to depict the capsule isolated from its environment. This also prevents the proposed 'build up' of energy in the bow of the moving system as proposed by some theorists. That is, we are replacing the 'flat' interior with an interior that encapsulates the engine completely isolating it from local or global space-time. That is, space-time is bending around the vessel, which includes space-time occupied by matter. Discussions of mass-energy buildup along the bow of such a superluminal vessel do not apply to a vessel that is isolated from normal space-time.

The number of measurements that must be taken just along the very finely sliced cross section of the forward edge in order to 'paint' just that very finely sliced cross section is very, very large; *and each must be taken at an extremely high rate of speed, for just that one cross sectional slice.* In order to 'paint' the entire manifold across a volume of space, a huge number of cross sectional slices must be measured in this way.

The energy requirement is minimal, merely enough to 'take the measurement.' The technology to 'take the measurements' at that rate is phenomenal, but within our reach. The phased array allows for the 'pointing' of the virtual photons that arise from the Near Field Effect or via 'real' impact by photons at an extremely high rate of speed and with extreme accuracy with an exacting determination of how long that photon will live, and take the measurement.

Keeping in mind, I lack the software to produce a better image; this is essentially, what we are measuring:

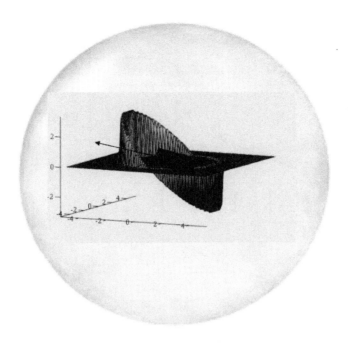

It must be noted here that, as stated, my depiction of the manifold is intentionally inverted from that which is considered in mainstream thinking.

The 'bubble' is a plasma of electron-positron pairs 'floating' on the surface of a static magnetic field, such as shown in nature of Earth's magnetosphere. Thus, for instance, the electrons would have migrated here to the top and the positrons to the bottom, depending on the orientation of the field. The two poorly drawn black rectangles are two phased array emitters (each emitter may have dozens of individual emitters, as drawn in Fig 1 and whose orientation is arbitrary), whose Near Field or 'real' photons are used. The arrow represents a single measurement being taken in the positive 'x' direction, keeping in mind that it is necessary to take measurements in all directions along the plane. I think it is worth mentioning that 'painting' the front and rear peaks is relatively easy, but the graduation from the front to the rear of the manifold by the slope between them is a challenge. The need to take measurements in the positive and negative 'y' axis (above and below) has to be determined experimentally. Speculating on this aspect of the measurement can go both ways and simply has to be derived experimentally.

Taking another visual look at the emitter/detectors:

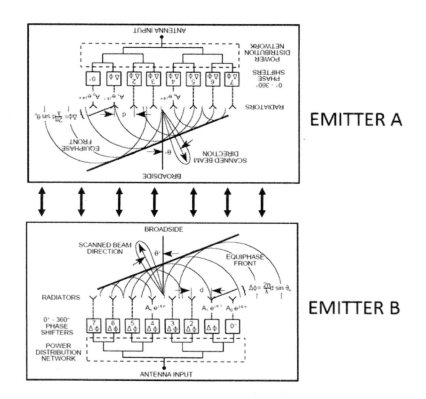

The plasma is held in a uniform state via an inhomogeneous magnetic field. The outward force is produced by their velocity from the emission, the inward force maintained by the magnetic field. As emissions from the phased array system interact with the plasma of electrons a secondary Compton wavelength of unique wavelength is emitted. Since the procession of the electron according to its spin state in the magnetic field will result in a slightly different velocity with respect to a stationary point as measured from the center of the magnetic field, the emitted Compton wavelengths for the two spin states will vary slightly from one another, giving a characteristic signature regarding the spin state of the electron the lambda has interacted with.

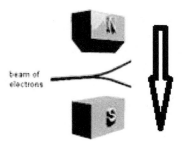

In this case, the lambda emission from the phased array is perpendicular to the path of the electrons, resulting in a different velocity and thus a different Compton wavelength for each electron that it has interacted with. In this case, the electron (or positron) has been both detected and its spin state in the plasma measured. The actual density of the plasma (luminosity) will have to be determined experimentally.

As depicted here, Compton scattering is a simple phenomenon to measure and gives characteristic wavelengths upon interacting with electrons.

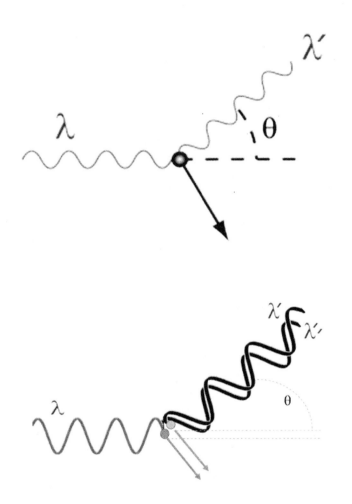

A photon of wavelength λ comes in from the left, collides with a target at rest, or in this case, motion that is to be detected, and a new photon of wavelength λ' emerges at an angle θ. In the case of measuring a second particle of differing spin state, the position of particle 2 will be slightly offset from particle one, yielding a lambda''. The phase difference between the two expectation values yields the spin state result. The slight phase difference between the two measurements tells us the path the electron is taking in the magnetic field and yields therefore the spin state of the particle.

A variation on this method that simplifies the approach and reduces computing power and increases measurement speed involves determining where the particle, spin +1/2 or -1/2 will be after passing through the magnetic field:

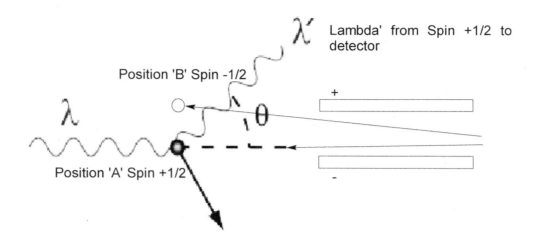

In this example the beam of electrons pass through the Stern-Gerlach field, I have arbitrarily assigned Spin +1/2 as the target for the incoming photon, which, if present at the predicted location emits a secondary photon that is detected. If no secondary photon is detected the particle is in Position 'B,' which in this scenario is Spin -1/2. The absence of a signal from the secondary photon indicates that the spin state is -1/2. The presence of a secondary photon in this scenario determines the particle to be spin +1/2, and the absence of a secondary photon indicates the spin state to be -1/2. Again, these assignments in this diagram are arbitrary and for demonstration purpose only.

This arrangement simplifies the system and increases detection speed and reduces computing power significantly. We are not measuring or computing redshift in this case, just the presence or absence of signal.

The entire arrangement requires speed and accuracy of the particle stream, predicting the particle's path according to its spin state after passing through the magnetic field, and extreme accuracy and precision in the placement of the primary photon. In summary, a stream of electron-positron pairs are created via a classic 'gold foil' type setup from a radioactive source:

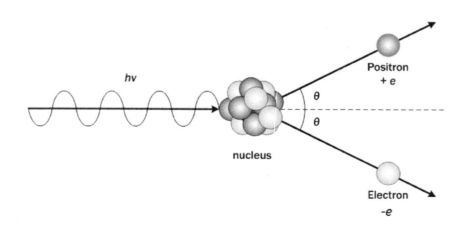

The particle stream is 'painted' around the engine source in the form of a spherical volume encompassing the engine via a system similar to a standard CRT arrangement:

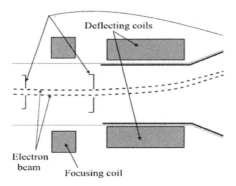

This system is used to 'paint' a sphere of finely separated electrons:

One particle stream passes through the Stern-Gerlach *filter: (note that both positions 'A' and 'B' are on the surface of the sphere)*

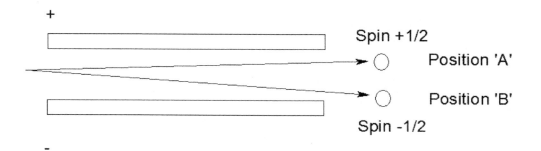

The emitted photons are controlled by a modified phased array system designed to emit the shortest possible wavelength in order to produce accuracy and precision.

The predicted position of, in this arbitrary example, Spin +1/2 is *probed* with an incoming photon in order to detect rather or not the particle is in Position 'A:'

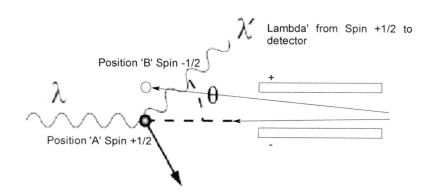

If the particle is in Position 'A' a secondary photon, lambda' is detected. If lambda' is not detected, in this scenario, the particle's spin state is -1/2. In this way, the particle's spin state is rapidly determined.

The rate of detection, R, as discussed (having been determined experimentally), determines t', which, obeying the laws of General Relativity *must* alter the local shape of space curvature, and is mapped into Alcubierre's original equation to 'paint' a space-time manifold.

The detection rate 'R' is mapped into Alcubierre's original equation, shown here in its quantized form, eliminating 'pi' and replacing 'pi' with an integer value of '3':

$$f(+/-)k \mapsto \frac{(\frac{Lp^4}{t_p^4})}{24G'} \frac{\left(\frac{\frac{nLp}{xt_p}}{\frac{1Lp}{1t_p}}\right)^2 (Lp_y^2 + Lp_z^2)}{4g'^2[(Lp_x - Lp_{xs}(t'))^2 + Lp_y^2 + Lp_z^2]} \left(\frac{df}{dr}\right)^2$$

The result is that the sphere is redefined by the local curvature of space-time, taking the form of Alcubierre's original space-time manifold:

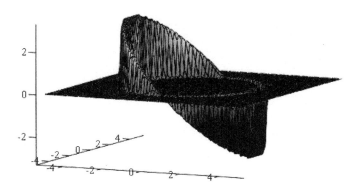

Since only photon emmision via the phased array system and detection are required the energy requirements to produce this space-time manifold are minimal, in the killowatt range, and requires no forms of 'exotic' energy or 'exotic' matter to produce.

PROOF OF PRINCIPLE

Proof of principle is achieved by pumping a stream of single electron-positron pairs in opposite directions, individually, like bullets. There is a principle here that is in line with Newton's 'for every action there is an equal and opposite reaction.' In nature, we look at the information above regarding length *dilation* in a gravity well. The only experimental evidence regarding what length does in a gravity well is that which is referenced above, and clearly indicates that length dilates in a gravity well.

It is a careful use of words here to state that although space is curving inward $(Lp \to 0)$, the 'objects' and things that fill that space *appear to dilate* with respect to their length $(Lp \to \infty)$. This *can be thought of as* a fixed length of an object crossing a greater number of Planck lengths as a result of local space curving inward $(Lp \to 0)$.

The notion that 'gravity holds space together' is a misdirected speculation of the observation of gravity holding the things in space together, not space itself. This is an obvious result of the fact that we have only speculative means for measuring the shape of space-time; gravitational lensing and so on. The rules change significantly on a microscopic scale. To date, no one has proposed a mechanism by which to measure the shape of space-time on a microscopic scale, only speculation that it can be achieved using laser interferometry. In this setup (for proof of principle), laser interferometry is used to determine if laser interferometry is suitable as a means of measuring the shape of space-time on a microscopic scale. At this time, we do not know what the scale will be.

The entire setup, positron-electron emission, Stern-Gerlach filter, and variable speed detector are set up on a bench top scale in one dimension. An attempt at producing the QZE and QAZE is researched and developed. At the time when we are confident we have achieved a QZE, we use a laser interferometer to measure any change in the quantitative length of space, L_p, otherwise known as L' in relativistic parlance.

The deviation of the red-shift from the predicted value is a confirmation that the shape of space has been altered, causing the photon to have a different wavelength than predicted – travelling a distance and time other than predicted.

As the rate of measurements is increased, the deviation from the predicted value will also increase. The means of determining if this is an artifact of the measurement or an actual shift in the shape of space-time is that the accuracy of the measurement will deviate but *not the precision.* That is, laser interferometry may not be appropriate on a bench top proof of principle scale. However, striking an electron in its now curved path that deviates from normal due to a change in the curvature of space at repeated intervals in its path with photons and measuring the phase shift in the Compton photons is a more direct solution to measuring the artificial curvature of space-time on a small scale.

That is, a laser interferometer is subject also to the curvature of space-time. For instance, Gravitational Lensing can be used to determine curved space, but not interferometry, as demonstrated by the failed Michelson-Morley experiment.

Thus, an approach similar to that of Gravitational Lensing, on a microscopic scale is my suggestion. By shooting a beam of coherent light across the QZE, and measuring its position at a target, we can determine space curvature. This is exactly how Arthur Eddington confirmed General Relativity in 1919 during a Solar Eclipse, by measuring the skewed positions of a few stars very near the sun's gravitational influence.

APPROACH II

In this second approach, we take a polymer capable of holding a charge (plasma) of electrons and bond it to a surface that is entirely composed of micro detectors. Each micro detector consists of an elevated P-N junction of sufficient static magnetic strength to deflect the path of an electron according to its spin state. At the base of the micro detector is a simple photoelectric effect type detector that determines which path the electron took, striking either point 'A' or point 'B.'

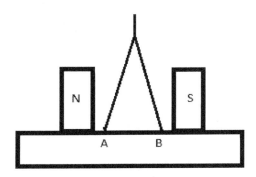

The outer surface is an appropriate polymer capable of bearing a current but not a conductor. The inner surface is composed of billions of these microprocessors each capable of taking millions of data points per second. The space-time manifold is controlled by the data acquisition rate across the surface of the object.

In both approaches it is noted that the Quantum Zeno Effect and QAZE is only applicable to quantum systems that change with time, such as radioactive decay and so on, which are well documented. In this case, by attempting to measure the spin state of the electron and/or positron we have to do so by measuring a change in position. By measuring the change in position, we acquire the knowledge of the spin state of the lepton. The measurement problem is averted as the spin states are not compromised by the rapidity of the measurements. Since we are not attempting to establish any significance to Quantum Entanglement or Superposition, the measurement problem does not exist as we are not making any attempt to glean such information out of the measurement, only the spin states.

By measuring the change in position at a phenomenal rate, we never actually or perhaps very slowly attain the knowledge of the spin state of the lepton. Therefore, this approach qualifies as an attempt to take advantage of the Quantum Zeno Effect in order to affect the progression of unitary time, and determine if the change in unitary time follows rules similar to those described by Special and General Relativity. The change in unitary time and the consequent effects on the shape of local space-time do not have to follow the same exact mathematical models and predictions according to Special and General Relativity, only the desired effect of a measurable change in the local shape of space-time is necessary. The actual change in the shape of local space-time will probably follow somewhat different rules and equations that will have to be derived experimentally and cannot at this time be predicted.

That is, covering a ship with these will not produce a large space-time manifold. However, as a gravitic engine, will provide unlimited lifting power.

It is more likely that Approach II can be used as a model for an engine that will work at low velocity in an atmosphere because the individual microprocessors are isolated from the environment, and Approach I is used at extreme high velocity in the vacuum of space. It would be practical as a 'lifting mechanism' negating the use of conventional chemical rockets, possessing extreme lifting power at low velocity to achieve reaching the vacuum of space, at which point the main engine can be brought on-line.

The type II engine is exactly what is needed to build an infrastructure of helium-3 mining operations on the lunar surface, as mass is no limitation. I actually consider the type II engine to be our priority at this time for this reason.

Raphael